HOME LINKS

Everyday Mathematics®

The University of Chicago School Mathematics Project

Mc Graw Hill Education

Bothell, WA • Chicago, IL • Columbus, OH • New York, NY

The University of Chicago School Mathematics Project

Max Bell, Director, *Everyday Mathematics* First Edition; James McBride, Director, *Everyday Mathematics* Second Edition; Andy Isaacs, Director, *Everyday Mathematics* Third, CCSS, and Fourth Editions; Amy Dillard, Associate Director, *Everyday Mathematics* Third Edition; Rachel Malpass McCall, Associate Director, *Everyday Mathematics* CCSS and Fourth Editions; Mary Ellen Dairyko, Associate Director, *Everyday Mathematics* Fourth Edition

Authors
Jean Bell, Max Bell, John Bretzlauf, Amy Dillard, Robert Hartfield, Andy Isaacs, James McBride, Rachel Malpass McCall, Kathleen Pitvorec, Peter Saecker

Fourth Edition Grade 1 Team Leader
Rachel Malpass McCall

Writers
Meg Schleppenbach Bates, Kate Berlin, Sarah R. Burns, Gina Garza-Kling, Linda M. Sims

Open Response Team
Catherine R. Kelso, Leader; Kathryn M. Rich

Differentiation Team
Ava Belisle-Chatterjee, Leader; Anne Sommers

Digital Development Team
Carla Agard-Strickland, Leader; John Benson, Gregory Berns-Leone, Juan Camilo Acevedo

Virtual Learning Community
Meg Schleppenbach Bates, Cheryl G. Moran, Margaret Sharkey

Technical Art
Diana Barrie, Senior Artist; Cherry Inthalangsy

UCSMP Editorial
Lila K.S. Goldstein, Senior Editor; Rachel Jacobs, Kristen Pasmore, Delna Weil

Field Test Coordination
Denise A. Porter

Field Test Teachers
Mary Alice Acton, Katrina Brown, Pamela A. Chambers, Erica Emmendorfer, Lara Galicia, Heather A. Hall, Jeewon Kim, Nicole M. Kirby, Vicky Kudwa, Stephanie Merkle, Sarah Orlowski, Jenny Pfeiffer, LeAnita Randolph, Jan Rodgers, Mindy Smith, Kellie Washington

Contributors
William B. Baker, John Benson, Jeanine O'Nan Brownell, Andrea Cocke, Jeanne Mills DiDomenico, Rossita Fernando, James Flanders, Lila K.S. Goldstein, Allison M. Greer, Brooke A. North, Penny Williams

Center for Elementary Mathematics and Science Education Administration
Martin Gartzman, Executive Director; Meri B. Forhan, Jose J. Fragoso, Jr., Regina Littleton, Laurie K. Thrasher

External Reviewers
The *Everyday Mathematics* authors gratefully acknowledge the work of the many scholars and teachers who reviewed plans for this edition. All decisions regarding the content and pedagogy of *Everyday Mathematics* were made by the authors and do not necessarily reflect the views of those listed below.

Elizabeth Babcock, California Academy of Sciences; Arthur J. Baroody, University of Illinois at Urbana-Champaign and University of Denver; Dawn Berk, University of Delaware; Diane J. Briars, Pittsburgh, Pennsylvania; Kathryn B. Chval, University of Missouri–Columbia; Kathleen Cramer, University of Minnesota; Ethan Danahy, Tufts University; Tom de Boor, Grunwald Associates; Louis V. DiBello, University of Illinois at Chicago; Corey Drake, Michigan State University; David Foster, Silicon Valley Mathematics Initiative; Funda Gönülateş, Michigan State University; M. Kathleen Heid, Pennsylvania State University; Natalie Jakucyn, Glenbrook South High School, Glenview, IL; Richard G. Kron, University of Chicago; Richard Lehrer, Vanderbilt University; Susan C. Levine, University of Chicago; Lorraine M. Males, University of Nebraska-Lincoln; Dr. George Mehler, Temple University and Central Bucks School District, Pennsylvania; Kenny Huy Nguyen, North Carolina State University; Mark Oreglia, University of Chicago; Sandra Overcash, Virginia Beach City Public Schools, Virginia; Raedy M. Ping, University of Chicago; Kevin L. Polk, Aveniros LLC; Sarah R. Powell, University of Texas at Austin; Janine T. Remillard, University of Pennsylvania; John P. Smith III, Michigan State University; Mary Kay Stein, University of Pittsburgh; Dale Truding, Arlington Heights District 25, Arlington Heights, Illinois; Judith S. Zawojewski, Illinois Institute of Technology

Note
Too many people have contributed to earlier editions of *Everyday Mathematics* to be listed here. Title and copyright pages for earlier editions can be found at http://everydaymath.uchicago.edu/about/ucsmp-cemse/.

www.everydaymath.com

Send all inquiries to:
McGraw-Hill Education
STEM Learning Solutions Center
8787 Orion Place
Columbus, OH 43240

ISBN: 978-0-02-137958-3
MHID: 0-02-137958-0

Printed in the United States of America.

1 2 3 4 5 6 7 8 9 RHR 19 18 17 16 15 14

Contents

Introducing *First Grade Everyday Mathematics*

Welcome to *First Grade Everyday Mathematics*. This program is a part of an elementary school mathematics curriculum developed by the University of Chicago School Mathematics Project.

Here are some features of the *First Grade Everyday Mathematics* program:

Children learn basic skills by solving problems based on everyday situations. They connect their own knowledge to their experiences both within and outside of school. Through these meaningful situations, children learn basic skills as mathematics becomes "real."

Children practice basic skills in a variety of engaging ways. They complete daily practice covering a variety of topics, find patterns on the number line, work with addition and subtraction facts, and play games that are designed to develop basic skills.

Children revisit concepts over the course of the year. To improve the development of basic skills and concepts, children regularly revisit concepts and repeatedly practice skills that have been taught earlier. The lessons are designed to build on concepts and skills throughout the year instead of treating topics in isolated sections.

First Grade Everyday Mathematics emphasizes the following topics:

- **Operations and Algebraic Thinking**
  Representing and solving problems involving addition and subtraction; understanding and applying properties of operations and the relationship between addition and subtraction to these problems; adding and subtracting within 20; and working with addition and subtraction equations

- **Number and Operation in Base Ten**
  Extending the counting sequence; understanding place value; and using place-value understandings and properties of operations to add and subtract within 100

- **Measurement and Data**
  Measuring lengths; telling and writing time; and representing and interpreting data

- **Geometry**
  Reasoning with shapes and their features

You will be provided with many opportunities to monitor your child's progress and to participate in your child's mathematics experiences. Throughout the year, you will receive Family Letters to keep you informed of the mathematical content your child will be studying in each unit.

Enjoy seeing your child's understanding of math grow as he or she connects mathematics to everyday life.

We look forward to an exciting year!

Counting

You will receive a Family Letter before each unit begins. Each letter introduces you to the content of the next unit, in this case, counting. The letter also includes vocabulary terms, activities you can do at home, descriptions of math games, and answers to the Home Links, or homework.

Unit 1 builds on what children learned about numbers in Kindergarten. In this unit, they review and practice counting. Children practice *rote counting*, or reciting numbers in order by 1s, 5s, and 10s. Children also practice *rational counting*, or counting collections of actual objects. After some experience, they begin to associate counting "1 more" or "1 less" with addition and subtraction. Children also use their counting skills to collect and record data using tally charts.

Number stories are also introduced in Unit 1. *Number story* is another name for what is sometimes called a "story problem" or a "word problem." Throughout *Everyday Mathematics,* number stories provide opportunities for children to use a variety of strategies to solve problems. Children are encouraged to talk through solving the number stories. Not only do they have many opportunities to solve number stories throughout first grade, but they are also asked to make up their own number stories.

Unit 1 introduces some of the tools used in *Everyday Mathematics,* such as pennies, dice, the Pattern-Block Template, pattern blocks, base-10 blocks, and the geoboard. Children also learn to navigate the number grid and use it to count by 1s and 10s.

Vocabulary
These are important terms your child learns in Unit 1. Listen to your child use these terms when talking about mathematics at home.

number grid A table in which numbers are arranged in order, usually 10 columns per row. A move from one number to the next within a *row* is a change of 1; a move from one number to the next within a *column* is a change of 10.

									0
1	2	3	4	5	6	7	8	9	10
11	12	13	14	15	16	17	18	19	20
21	22	23	24	25	26	27	28	29	30
31	32	33	34	35	36	37	38	39	40
41	42	43	44	45	46	47	48	49	50

number line A line with numbers that are marked in order.

1   2   3   4   5   6   7   8   9   10

number story A story that involves numbers and one or more questions. For example, *I have 7 crayons. Carrie gave me 5 more crayons. How many crayons do I have now?*

tally chart A chart that uses tally marks to track values in a set of data.

Number of Pull-Ups	Number of Children
0	~~HHT~~ /
1	~~HHT~~
2	////
3	//

tally mark A mark used in a count. Tally marks let children represent numbers they can count and say, but may not be able to write yet.

~~HHT~~ ///

toolkit Individual bags or boxes used in the classroom; they usually contain a variety of items—such as calculators, measuring tools, and manipulatives—which help children understand mathematical ideas.

Do-Anytime Activities

To work with your child on concepts taught in this unit, try these activities:

- Discuss examples of mathematics in everyday life: TV listings, road signs, recipe measurements, time, and so on.

- Count orally by 5s and 10s when doing chores or riding in the car or on a bus. Occasionally count down, or back; for example: 90, 80, 70, 60, and so on.

- Count numbers of objects around the house and while shopping. Have your child keep track using tally marks. For example, count the number of canned goods bought at the grocery store.

Building Skills through Games

Your child will play these games In Unit 1:

Bunny Hop
Players roll a die to navigate on a number line to 20 and back to 0.

Monster Squeeze
The leader chooses a mystery number on a number line. Other players try to guess the number using clues from the leader.

Penny-Dice
Players take turns rolling a die and taking the number of pennies indicated on the die. The first player to get 20 pennies wins.

Rolling for 50
Players roll a die to navigate on the number grid. The first player to reach FINISH wins.

Top-It
Each player turns over a number card from a deck. Whoever has the higher number keeps both cards. Whoever has more cards when the whole deck has been used wins.

As You Help Your Child with Homework

Your child will bring home assignments called "Home Links." Home Links are suggested follow-up or enrichment activities to be done at home. They will not take much time to complete, but may involve interaction with an adult or an older child. Each Home Link activity is identified by the following symbol:

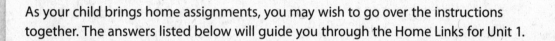

As your child brings home assignments, you may wish to go over the instructions together. The answers listed below will guide you through the Home Links for Unit 1.

Home Link 1-7

1. Your child should attach pictures of numbers as they appear in everyday life.

Home Link 1-8

1. Answers vary.

2. 1; 2; 4; 6; 8; 9

Home Link 1-9

1. Your child may mention pattern blocks, base-10 blocks, or geoboards.

2. 7

Home Link 1-10

1. Sample number story: There are 5 flowers in the garden. If I pick 1 of them to give to my teacher, how many flowers will be left?
Answer: 4 flowers

NOTE: Encourage your child to come up with his or her own way to solve the problem, whether it's drawing pictures or counting on fingers. As an adult you know that $5 - 1 = 4$, but it is more natural for your child to come up with his or her own strategy than to think of the number story as $5 - 1 = 4$.

Your child should attach the picture used for the number story to the page if he or she didn't already draw it.

2. 4, 7, 11

Home Link 1-11

1. Check that your child can count by 1s to the number he or she wrote.

2. Sample answer: 50, 40, 30, 20, 10, 0

3. Sample answer: I can count squares from left to right as I count by 1s. To count by 10s, I can start at the top right corner and move down.

4. 15; 20; 25; 35; 40; 50

Open Response and Reengagement Lessons

A two-day lesson in each unit of *First Grade Everyday Mathematics* is an Open Response and Reengagement lesson. In these lessons, children solve interesting problems using their own strategies and reasoning. On Day 1, children solve an open response problem—a problem with more than one possible strategy or solution. On Day 2, the class discusses children's work from Day 1 to "reengage" with the problem and learn more about the mathematics involved. Children then revise their work based on what they learn from the discussion.

These lessons are not assessments, but opportunities for children to solve approachable problems that require persistence. Children's work on Day 1 reveals both strengths and weaknesses, allowing the second day's discussion to focus on areas that need improvement. From these discussions, children find that learning from mistakes is a natural part of mathematical problem solving. Explaining their thinking and listening to the explanations of others builds children's confidence while allowing them to see that there is more than one way to solve a problem. This promotes creative thinking about solutions later on. Having an opportunity to revise their work helps children realize that they can be successful tackling hard tasks if they think about them and keep trying.

The open response problem in this unit asks children to count a group of objects and choose strategies, such as grouping by 2s or 5s, to ensure that they count accurately and efficiently.

Drawing of child's strategy for counting by 5s

These lessons continue work on problem solving that is central to *Everyday Mathematics* across all the grades. Ask your child to talk to you about the problems and his or her mathematical thinking throughout the year. Enjoy seeing your child become a confident problem solver!

Numbers Are Everywhere

Family Note

As mentioned in a previous Family Letter, your child will have Home Link assignments throughout the year. This is your child's first Home Link. Home Links appear in the first-grade program for many reasons:

- The assignments encourage children to take initiative and responsibility. As you respond with encouragement and assistance, you help your child build independence and self-confidence.

- Home Links reinforce newly learned skills and concepts. They provide opportunities for your child to think and practice at his or her own pace.

- These assignments relate the mathematics your child is learning in school to the real world, which is very important in the *Everyday Mathematics* program.

- Home Links will give you a better idea of the mathematics your child is learning.

Listen and respond to your child's comments about mathematics. Point out examples of numbers (time, TV channels, page numbers, telephone numbers, bus routes, lists, and so on). Children who do math with someone learn math. For this reason, *Everyday Mathematics* provides many counting and thinking games that you and your child will have fun playing together and that will help build a strong understanding of mathematics.

For this first Home Link, your child might look for a newspaper ad for grocery items, a calendar page, or a picture of a clock. This activity helps expand your child's awareness of numbers in the world.

Please return this Home Link to school tomorrow.

Cut examples of numbers from scrap papers you find at home.

Glue some examples on the back of this page.

You can also bring examples that will not fit on this page to school.

Do not bring anything valuable!

Organizing Data with Tally Marks

Family Note

Today, your child used tally marks as the class collected data by counting. Tally marks let children represent numbers they can count and say, but may not yet be able to write, and they are useful for keeping track of data collected by counting. Remind your child that the fifth tally mark crosses the other four, like this: ‖‖ Encourage your child to first count by 5s for groups of 5 tallies and then count by 1s. For example, ‖‖ ‖‖ ‖‖ ||| should be counted as 5, 10, 15, 16, 17, 18. Developing this skill will take some practice.

Please return this Home Link to school tomorrow.

(1) Write 6 numbers. Make tally marks for each number.

Number	Tally Marks
18	‖‖ ‖‖ ‖‖ |||

Practice

(2) Count. Write the missing numbers.

−1 0 ___ ___ 3 ___ 5 ___ 7 ___ ___ 10 11 12 13 14 15 16 17 18 19

eleven   11

Exploring Math Materials

Family Note

In *First Grade Everyday Mathematics*, children regularly engage in Exploration activities. These activities provide children with hands-on experiences using classroom tools, collecting data, solving problems, and playing math games. During Exploration days, children rotate through different stations in small groups, focusing on a new activity at each station. These stations give each child the opportunity to participate in several activities during math class. Please ask your child about today's mathematics Explorations that included using base-10 blocks, pattern blocks, and geoboards.

Please return this Home Link to school tomorrow.

① Tell someone at home about your favorite mathematics Exploration.

Draw something you did in your Explorations today.

Practice

② How many dots? _____ dots

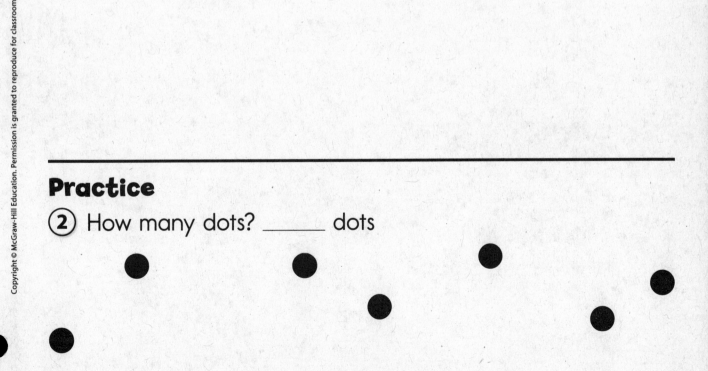

Number Stories

Family Note

Number story is another name for what is sometimes called a "story problem" or a "word problem." *Everyday Mathematics* uses *number story* to emphasize that the story must involve numbers. Help your child illustrate one below.

Please return this Home Link to school tomorrow.

(1) Find or draw a picture of a group of things, such as animals, people, flowers, or toys.

Tell a number story about your picture to someone at home.

Then attach your picture to this page.

Practice

(2) Write each number.

///// _____

卌 // _____

卌 卌 / _____

Counting Up and Back

Family Note

Today your child used the number grid for counting larger numbers. Notice the different ways you can count on it. Move to the right within a row to count by 1s. Move down in the same column to count up 10s. Count with your child with and without the number grid. Listen as your child counts by 1s and 10s. Counting aloud for someone else provides good practice for this essential first-grade skill.

Please return this Home Link to school tomorrow.

(1) Count up by 1s, starting with 1. I counted to _____.

(2) Count back by 10s. Start with 50 or the highest number you can. I started with _____.

(3) Explain to someone at home how to use the number grid to help with counting.

									0
1	2	3	4	5	6	7	8	9	10
11	12	13	14	15	16	17	18	19	20
21	22	23	24	25	26	27	28	29	30
31	32	33	34	35	36	37	38	39	40
41	42	43	44	45	46	47	48	49	50
51	52	53	54	55	56	57	58	59	60
61	62	63	64	65	66	67	68	69	70
71	72	73	74	75	76	77	78	79	80
81	82	83	84	85	86	87	88	89	90
91	92	93	94	95	96	97	98	99	100
101	102	103	104	105	106	107	108	109	110
111	112	113	114	115	116	117	118	119	120

Practice

(4) Count up by 5s.

5, 10, _____, _____, _____, 30, _____, _____, 45, _____

Introducing Addition

In Unit 2, your child begins learning strategies for solving addition problems. Children create a class "Strategy Wall" that lists all the strategies they learn and practice. Strategies covered in this unit include counting on, using the turn-around rule, and using pairs of numbers that add to 10 (such as 3 and 7, or 9 and 1). Children will continue to learn strategies that help them become fluent with addition within 20 as the year progresses.

An important tool for addition is the ten frame. Ten frames are especially helpful for identifying pairs of numbers that add to 10, as well as for for illustrating other facts within 10.

10 frames: 7 dots and 3 blanks

$7 + 3 = 10$

7 dots: 1 full column of 5 dots and 1 column with 2 dots

$5 + 2 = 7$

Children also begin modeling number stories using change diagrams to organize information. (*See below.*) They use numbers and symbols to write number models that represent these problems.

Vocabulary

change diagram A diagram used in *Everyday Mathematics* to model situations in which quantities are either increased or decreased. The diagram includes a starting quantity, an ending quantity, and the amount of change.

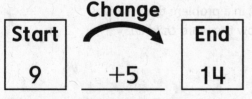

Start	Change	End
9	+5	14

A change diagram for $9 + 5 = 14$

counting on An addition strategy that involves starting with one number being added and counting on the other number. For example, to solve 5 + 3, start at 5 and count on.

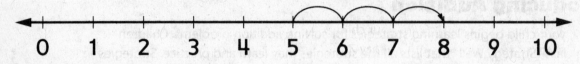

Math Boxes A collection of problems to practice skills.

ten frame An array of 10 squares used to organize small numbers.

Ten frame showing 6

turn-around rule A rule for solving addition problems based on a property of addition. If you know that 6 + 8 = 14, then, by the turn-around rule, you also know that 8 + 6 = 14.

unit box A box displaying the unit for numbers. For example, in a problem that involves the number of children in a classroom, the unit box would show the word *children*.

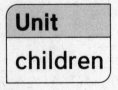

Do-Anytime Activities

To work with your child on concepts taught in this unit and in Unit 1, try these activities:

1. Select a number less than 10. Have your child name the other number needed to make a sum of 10. For example, if you say *7*, your child should say *3*.

2. Create number stories together and solve them using a change diagram or a number grid. For example, "Carrie had 14 stickers. She gave 3 of them to her friends. How many stickers does Carrie have now?"

−9	−8	−7	−6	−5	−4	−3	−2	−1	0
1	2	3	4	5	6	7	8	9	10
⑪	12	13	⑭	15	16	17	18	19	20
21	22	23	24	25	26	27	28	29	30
31	32	33	34	35	36	37	38	39	40
41	42	43	44	45	46	47	48	49	50

Counting back from 14

3. Make up number stories and number models together for everyday events. For example, when riding in the car, count things you see and make up stories such as: "I saw 3 red cars. Then I saw 2 blue cars. How many cars did I see in all? $3 + 2 = ?$"

Building Skills through Games

Your child will play these games and others in Unit 2.

High Roller

Players roll two dice. They keep the die with the greater number (the high roll) and then reroll the other die. They count on from the high roll to get the sum of the two dice.

Penny Plate

Players begin with a specified number of pennies, usually 10. One player hides some of the pennies under the plate. The other player counts the visible pennies and guesses how many pennies are hidden using knowledge of numbers that add to 10.

Roll and Total

Players roll a pair of dice: one dot die and one labeled with the numerals 3 through 8. They find the sum and record the result.

Ten-Frame Top-It

Children compare the numbers of dots on ten-frame cards in this variation of *Top-It*.

As You Help Your Child with Homework

As your child brings assignments home, you may want to go over the instructions together, clarifying them as necessary. The answers listed below will guide you through the Home Links for this unit.

Home Link 2-1

1. 8; 8

2. Explanations will vary but should include that the numbers are being added in a different order, but the answer is the same.

3. 30; 35; 40; 50

Home Link 2-2

1. Sample answers:

Number of Pennies in One Hand	Number of Pennies in the Other Hand
5	5
8	2
7	3
1	9

2. 6

Home Link 2-3

1. Answers vary.

2. 10

Home Link 2-4

1. 4, 6   2. 8, 2   3. 5, 5   4. Answers vary.

Home Link 2-5

1-2. Answers will vary but should show a total of 10 toys; some dolls and some blocks.

3. 5

Home Link 2-6

Sample answers given for 1 and 2.

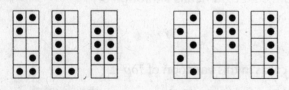

3. 40; 45; 55; 65

Home Link 2-7

1. Answers vary.   **2.** 5

Home Link 2-8

1. 5;

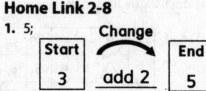

2. 9;

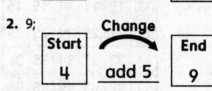

3. 20, 21, 22, 23, 24

Home Link 2-9

1.

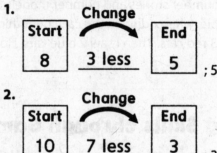
; 5

2.

Start 10   Change 7 less   End 3
; 3

3. 11   4. 15   5. 3

Home Link 2-10

1. $4 + 4 = 8$; 8

2. $9 - 3 = 6$; 6

3. Answers vary.

Home Link 2-11

1. $5 + 3 = \boxed{}$; 8

2. Sample answer: Sophie had 7 crayons. She lost some crayons. She has 3 crayons left. How many crayons did Sophie lose?

3. Sample answer: 3 and 7, 7 and 3

Introducing Addition Strategies

(1) Solve.

Jai has 3 shells.
His sister has 5 shells.
How many shells do they have in all?

_____ shells

Ellen found 5 rocks.
Her friend found 3 rocks.
How many rocks do they have in all?

_____ rocks

(2) Explain to someone at home how these number stories are alike and how they are different.

Practice

(3) Count up by 5s.

20, 25, _____, _____, _____, 45, _____

Two-Fisted Penny Addition

① Do Two-Fisted Penny Addition with someone at home:

- Place 10 pennies on the table. Grab some pennies with one hand. Pick up the rest with the other hand.

- Place each handful of pennies in its own pile.

- Use the table below to write how many pennies are in each pile.

Number of Pennies in One Hand	Number of Pennies in the Other Hand

Practice

② Shane has 4 model boats. He buys 2 more boats.

How many boats does Shane have now?

_____ boats

Pairs of Numbers That Add to 10

① Manny has 10 balloons.

Some of the balloons are blue.

Some of the balloons are yellow.

Draw a picture to show how many of each color balloons Manny might have.

Practice

② Start at 5. Count up 5. Where do you land? _____

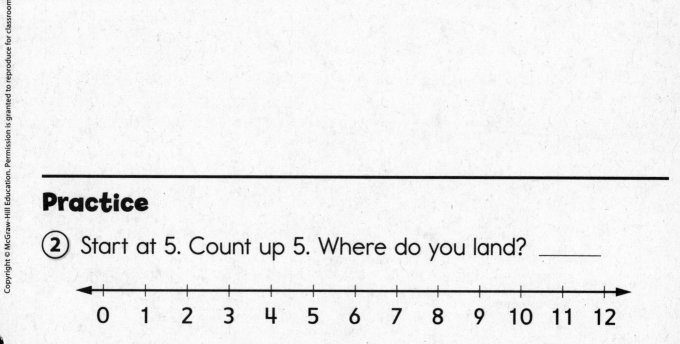

Family Letter

Quick Looks in *First Grade Everyday Mathematics*

Throughout first grade, children engage in activities referred to as "Quick Looks." Quick Looks use images of dot patterns or ten frames to encourage children to break numbers apart and put them together in flexible ways. Being able to think flexibly about numbers is an important skill to help children develop strategies for solving addition and subtraction facts. Children are shown each image for 2–3 seconds, and then they share *what* they saw and *how* they saw it.

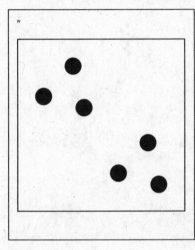 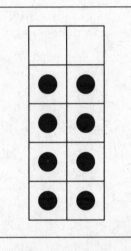 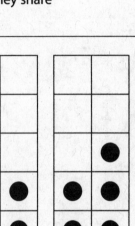 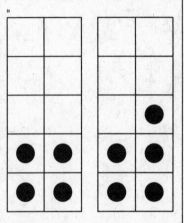

Sample Quick Look images

Children explain finding the total number of dots in the second image above in various ways, such as, "I saw 4 and 4 and that makes 8," "I skip counted: 2, 4, 6, 8," and "There are 2 missing from the ten frame, and I know $10 - 2 = 8$." Quick Looks with more complex images are used later to help children develop important fact strategies for solving more difficult facts. *See below:*

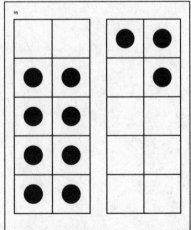

To solve $8 + 3$, children mentally manipulate the images to "make 10." So $8 + 3 = 10 + 1 = 11$.

Encourage your child to talk about the ways that he or she figured out the total dots on the Quick Looks done in class.

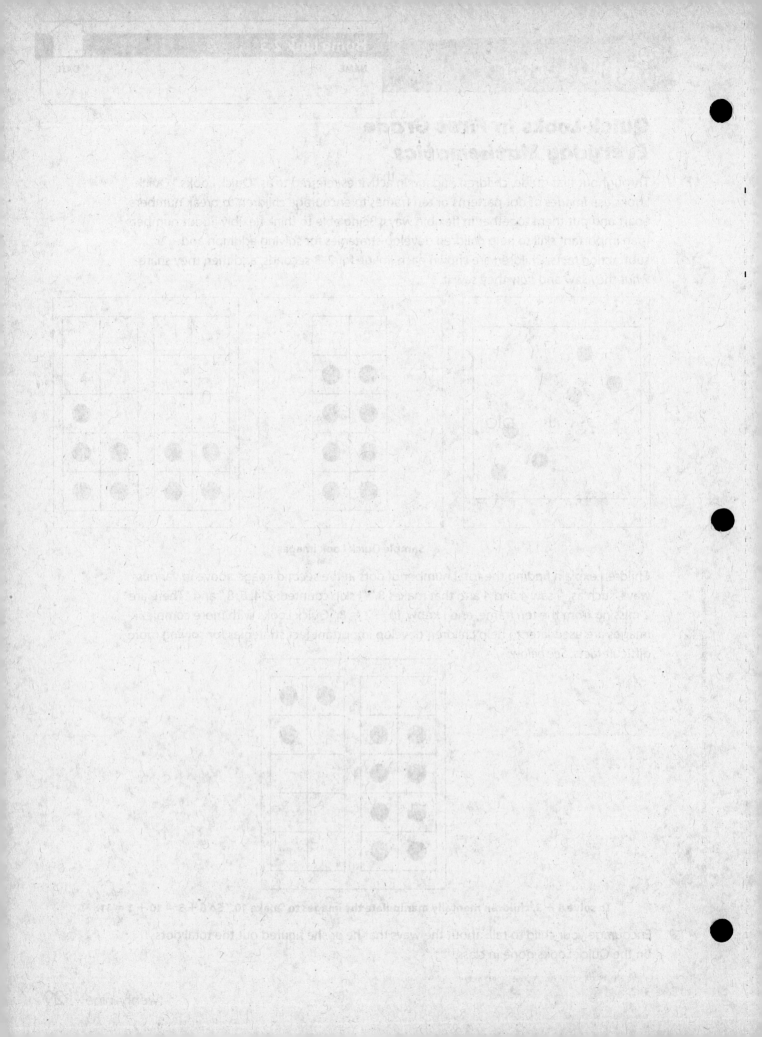

Ten Frames

Family Note

One way children explore pairs of numbers that add to 10 is by using ten frames. In one of today's Explorations, your child used a ten frame to help find all the possible combinations of 10 red and green apples. Also in today's lesson, children played a subtraction game and collected data about objects in the room.

Three ten frames are shown below. Have your child explain to you why these are called ten frames. Throughout the year, children will work with ten frames to help them identify, compare, break apart, and add numbers.

Please return this Home Link to school tomorrow.

① How many dots are in the ten frame? _____

How many blank spaces are in the ten frame? _____

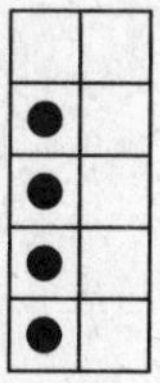

② How many dots are in the ten frame? _____

How many blank spaces are in the ten frame? _____

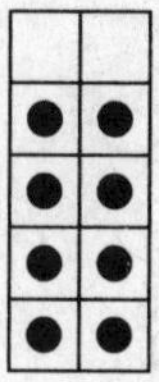

③ How many dots are in the ten frame? _____

How many blank spaces are in the ten frame? _____

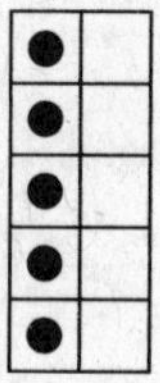

Practice

④ How many chairs are in your house?

_____ chairs

Finding All of the Ways to Make 10

(1) There are 10 toys in a toy box.
Some toys are blocks, and some toys are dolls.
Draw a picture of the toys in the toy box.

(2) Draw a different picture of toys in the toy box.

Practice

(3) There are 2 bananas and 3 oranges in a bowl.
How many pieces of fruit are there in all?

_____ pieces of fruit

More Uses of Ten Frames

(1) Show the number 6 in three different ways.

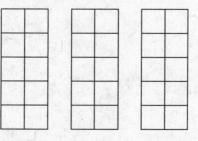

(2) Show the number 5 in three different ways.

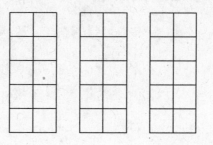

Practice

(3) Count by 5s.

35, _____, _____, 50, _____, 60, _____, 70

Labeling Counts

Family Note

In everyday life, numbers are used in contexts. You seldom encounter just the number 6. You see 6 cans of juice, 6 dollars, a length of 6 feet, and so on. In class your child will be asked to put numbers in context, too. The unit box is a reminder to children that they should consider the contexts of the numbers they are using.

Please return this Home Link to school tomorrow.

① Draw a picture of a group of objects.
Fill in the unit box.
Tell how many objects you counted.

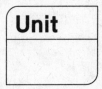

Unit

Practice

② Start at 9.
Count back to 4.
How many spaces did you move?
_____ spaces

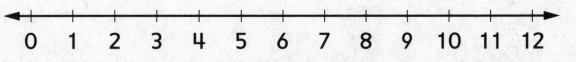

Change-to-More
Number Stories

Family Note

Your child first learned about number stories in Unit 1. Today children examined a specific type of
number story called change-to-more, in which more is added to the starting amount. To help make
sense of these situations, children use change diagrams (shown below) to help them organize the
information in the story.

Please return this Home Link to school tomorrow.

Solve. Fill in the change diagrams.

① Kendra hit 3 home runs at her first softball game.
She hit 2 home runs at her next game.
How many home runs did she hit in all? _____ home runs

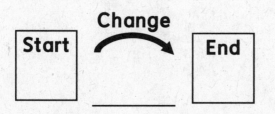

② David told 4 jokes yesterday.
He told 5 more jokes today.
How many jokes has David told in all? _____ jokes

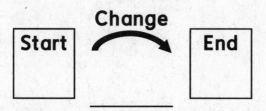

Practice

③ Count by 1s.

17, 18, 19, _____, _____, _____, _____, _____

Change-to-Less Number Stories

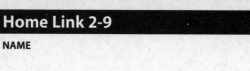

Copyright © McGraw-Hill Education. Permission is granted to reproduce for classroom use.

Family Note

Today your child continued exploring real-life number stories. Children learned about change-to-less number stories, in which things are taken away from starting quantities. Work with your child to make sure he or she can identify the information needed to complete the change diagrams below.

Please return this Home Link to school tomorrow.

Complete the change diagrams to solve the number stories.

① Erin had 8 balloons. 3 balloons popped.
How many balloons does Erin have left?

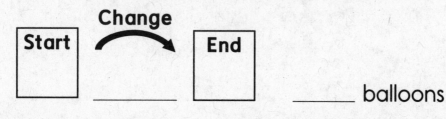

_____ balloons

② Kesha had 10 books. She read 7 of the books.
How many books does Kesha have left to read?

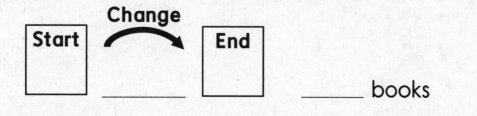

_____ books

Practice

How many tally marks?

③ ~~HHT~~ ~~HHT~~ I _____ tally marks

④ ~~HHT~~ ~~HHT~~ ~~HHT~~ _____ tally marks

⑤ ||| _____ tally marks

Number Models

Family Note

In the last two lessons, children worked with change-to-more and change-to-less situations using change diagrams. Today they wrote number models using numbers and mathematical symbols (+, −, =) to represent these number stories. Do not worry if your child still needs help writing number models. There will be many opportunities for your child to practice this throughout the year.

Please return this Home Link to school tomorrow.

Write a number model for each number story.
Use the change diagrams to help you.

(1) Rebecca read 4 books last week. She read 4 more books this week. How many books did Rebecca read in all?

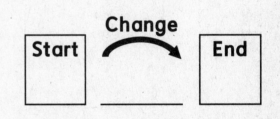

_____ + _____ = _____

_____ books

(2) The zoo had 9 lions. 3 lions moved to another zoo. How many lions were left?

_____ − _____ = _____

_____ lions

Practice

(3) Count the windows in your home. Use tallies to show how many windows you have.

_____ windows

Finding Unknowns

Family Note

Today your child wrote number models to represent number stories in which various pieces of information were missing or unknown. Children used an empty box like this ☐ to show which part was unknown. Once they solved the number story, they filled in the box with the correct answer. This skill will be further explored and practiced many times throughout the year.

Please return this Home Link to school tomorrow.

① Write a number model to describe the number story.
Then solve the problem.

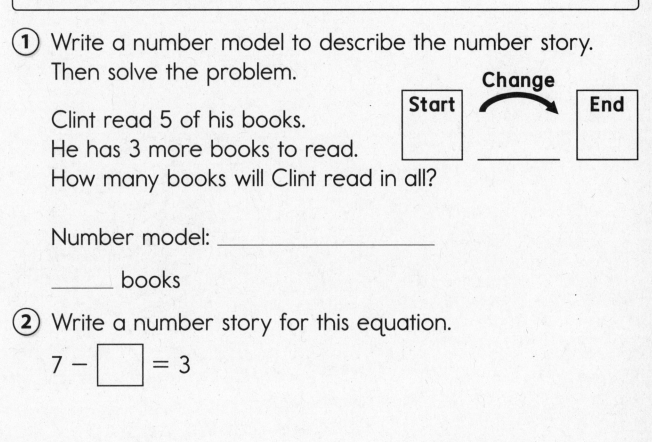

Clint read 5 of his books.
He has 3 more books to read.
How many books will Clint read in all?

Number model: _____

_____ books

② Write a number story for this equation.

$7 - \boxed{} = 3$

Practice

③ Write a pair of numbers that add to 10.
Use the turn-around rule to write another pair.

Unit 3: Family Letter

Number Stories

In Unit 3, children continue learning about number stories. In Unit 2, they solved stories based on change situations in which starting quantities get larger or smaller. In Unit 3, they learn to model and solve parts-and-total situations in which two quantities are combined to make larger quantities. They use parts-and-total diagrams to organize the information about these situations and write number models to describe them.

Total	
17	
Part	Part
9	8

Janis had 9 blue fish and 8 red fish.

How many fish does Janis have in all?

$$9 + 8 = 17$$

Children also explore the relationship between counting and addition and subtraction. They practice counting up and back on number lines to add and subtract, investigate patterns in number lines and number grids that will help them count more efficiently, and do activities with calculators and Frames-and–Arrows diagrams to help them connect counting patterns to addition and subtraction.

Vocabulary   Important terms in Unit 3:

Frames and Arrows A diagram consisting of frames connected by arrows used to represent number sequences. Each frame contains one number, and each arrow represents a rule that determines what number goes in the next frame.

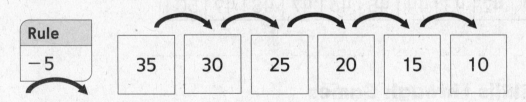

Rule
−5

| 35 | 30 | 25 | 20 | 15 | 10 |

The Family Note on Home Link 3-9, which you will receive later, provides a more detailed description of Frames and Arrows.

Parts-and-Total Diagram A diagram used to model problems in which two or more quantities (parts) are combined to get a total quantity.

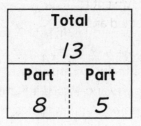

Total	
13	
Part	**Part**
8	*5*

Do-Anytime Activities

To work with your child on concepts taught in this and previous units, try these activities:

1. Count by 1s, 5s, or 10s aloud with your child. For a challenge, try starting at different numbers. "Count up by 10s starting at 3: 3, 13, 23, 33 …"

2. Have your child tell you a number story for a given number sentence, such as 3 + 5 = 8. For example, "I had 3 dogs. Then I got 5 more dogs. Now I have 8 dogs."

3. Using the number grid, select a number and have your child point to a number that is 1 more or 1 less than the selected number. Or do problems like this: "Start at 28. Count back (or up) 5 spaces. On which number did you land?"

−9	−8	−7	−6	−5	−4	−3	−2	−1	0
1	2	3	4	5	6	7	8	9	10
11	12	13	14	15	16	17	18	19	20
21	22	23	24	25	26	27	28	29	30
31	32	33	34	35	36	37	38	39	40
41	42	43	44	45	46	47	48	49	50

Building Skills through Games

Your child will play these games and others in Unit 3:

Domino Top-It

Children compare total numbers of dots on dominoes in this variation of *Top-It*.

Subtraction Bingo

Bingo cards have a number in each space. Players take turns flipping over two number cards and calling out the difference. They mark the differences on their cards until one player covers 4 spaces in a row.

As You Help Your Child with Homework

As your child brings home assignments, you may want to go over the instructions together, clarifying them as necessary. The selected answers listed below will help guide you through the Home Links for this unit.

Home Link 3-1

1. 8

2. 2

3. 45, 46, 47, 48

Home Link 3-2

1. 4; 3

2. Sample answer: Mark had 4 blue ribbons. He won 3 more blue ribbons. Now he has 7 blue ribbons.

3. 16

4. 20

5. 35

Home Link 3-3

1. Answers vary.

2. Answers vary.

3. 5; 5; Yes

Home Link 3-4

1. Pictures vary but should show 5 fish and 3 cats.

2. 8; Sample answer: 5 + 3 = 8

3. 7

Home Link 3-5

1. 10, 20

2. 5, 10, 15, 20

3. 2, 4, 6, 8

4. 1, 2, 3, 4

5. Sample answer: When you count by 5s, you land on all the count-by-10 numbers plus some more numbers.

6. 10

7. 8

Home Link 3-6

1. 13

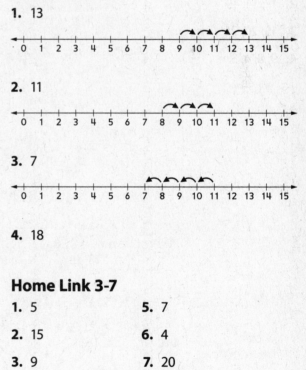

2. 11

3. 7

4. 18

Home Link 3-7

1. 5

2. 15

3. 9

4. 6

5. 7

6. 4

7. 20

8. ~~HHH~~ ~~HHH~~ ~~HHH~~ ~~HHH~~

Home Link 3-8

1.

									0
1	2	3	4	✗	6	7	8	9	✗
11	12	13	14	✗	16	17	18	19	✗
21	22	23	24	✗	26	27	28	29	✗
31	32	33	34	✗	36	37	38	39	✗
41	42	43	44	✗	46	47	48	49	✗

2. 10; 5

3. 24

4. 31

5. 39

6. 46

Home Link 3-9

1. 7; 11; 15

2. 17; 14; 13

3. 15; 20; 25

4. 9; 8; 4; 3

Home Link 3-10

1. Sample answers: add 2; + 2; count up by 2s

2. Sample answers: add 5; + 5; count up by 5s

3. Sample answers: subtract 3; − 3; count back by 3s

4. Answers vary.

Home Link 3-11

1. 3, 6, 9, 12, 15, 18, 21

2. 20, 18, 16, 14, 12, 10, 8

3. 45, 40, 35, 30, 25, 20, 15

4. 9 + 9 = 18; 7 + 3 = 10

Parts-and-Total Situations

Solve. Use the parts-and-total diagrams to help you.

① Jenny answered every test question.
She got 5 right and 3 wrong.
How many questions were on the test?

_____ questions

Total	
Part	**Part**

② There are 10 kids on Kevin's team.
8 are boys.
The rest are girls.
How many girls are on the team?

_____ girls

Total	
Part	**Part**

Practice

③ Count by 1s.

43, 44, _____, _____, _____, _____

Practicing Number Stories

(1) Solve. Explain your strategy to someone at home.
Walt was at the carnival.
He had 8 carnival tickets and 2 pens.
He traded 4 tickets for 1 more pen.

How many tickets does Walt have now? _____ tickets

How many pens does Walt have now? _____ pens

(2) Write a number story that matches this number sentence.
$4 + 3 = 7$

_____

_____

_____

Practice

Count up to solve.

(3) What number is 4 more than 12? _____

(4) What number is 3 more than 17? _____

(5) What number is 9 more than 26? _____

Ordering Objects by Length

Family Note

Today your child explored length by directly comparing the lengths of objects side-by-side. This activity helps prepare children to compare lengths of objects indirectly and to measure length more formally. In the next unit, your child will begin to measure length using items in the classroom, such as blocks and paper clips. Children also explored doubles facts and counted a collection of objects. Children also counted the number of chair legs in the classroom and wrote number models for matching pairs.

Please return this Home Link to school tomorrow.

1. Find two things in your home that are about the same length. Draw pictures of them.

2. Find two things that have very different lengths. Draw pictures of them.

Practice

3. Start at 2. Count up 3. Where do you land?

   Start at 3. Count up 2. Where do you land?

   _____   _____

   Did you end at the same number for both? _____

Number Stories

(1) Dillon's family has 5 fish and 3 cats.
Draw a picture of the fish and the cats.

How many pets does Dillon's family have in all?

_____ pets

Write a number model.

_____

Practice

(2) Laura had 14 tickets.
She bought an eraser.
She has 7 tickets left.
How many tickets did the eraser cost?

14 − _____ = 7

Unit
tickets

Counting on Number Lines

Family Note

Today your child learned many ways to count on a number line. Ask your child to tell you about patterns in number-line counts.

Please return this Home Link to school tomorrow.

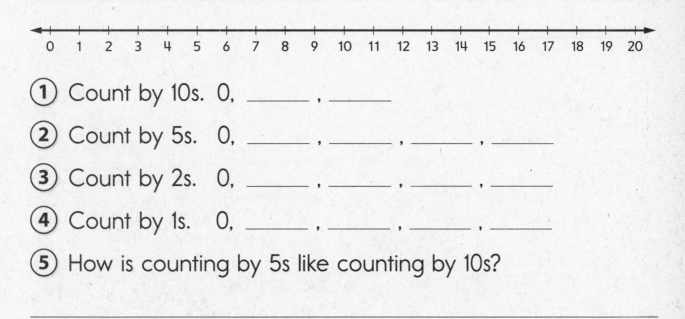

① Count by 10s. 0, _____ , _____

② Count by 5s. 0, _____ , _____ , _____ , _____

③ Count by 2s. 0, _____ , _____ , _____ , _____

④ Count by 1s. 0, _____ , _____ , _____ , _____

⑤ How is counting by 5s like counting by 10s?

_____

_____

_____

Practice

Find the sums.

⑥ 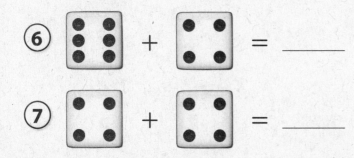 + = _____

⑦ + = _____

fifty-nine **59**

Counting to Add and Subtract

Family Note

Today your child solved problems like 3 + 2 and 8 − 5 by counting up and back on a number line. Ask your child to show you how to do this.

Please return this Home Link to school tomorrow.

Draw hops on the number line to help you solve these problems.

(1) 9 + 4 = _____

```
<──┼──┼──┼──┼──┼──┼──┼──┼──┼──┼──┼──┼──┼──┼──┼──┼──>
   0  1  2  3  4  5  6  7  8  9  10 11 12 13 14 15
```

(2) 3 + 8 = _____

```
<──┼──┼──┼──┼──┼──┼──┼──┼──┼──┼──┼──┼──┼──┼──┼──┼──>
   0  1  2  3  4  5  6  7  8  9  10 11 12 13 14 15
```

(3) 11 − 4 = _____

```
<──┼──┼──┼──┼──┼──┼──┼──┼──┼──┼──┼──┼──┼──┼──┼──┼──>
   0  1  2  3  4  5  6  7  8  9  10 11 12 13 14 15
```

Practice

(4) Circle the winning card in *Top-It*.

More Counting to Add and Subtract

Find the missing number.
Use the number line.

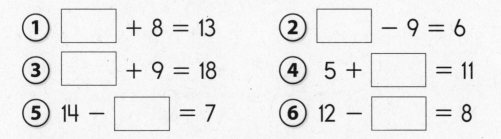

① ☐ $+ 8 = 13$ ② ☐ $- 9 = 6$

③ ☐ $+ 9 = 18$ ④ $5 +$ ☐ $= 11$

⑤ $14 -$ ☐ $= 7$ ⑥ $12 -$ ☐ $= 8$

Practice

Flowers	
Roses	卌 卌 ||
Daisies	卌 |||

⑦ How many flowers in all?

_____ flowers

⑧ Make tally marks to show how many flowers in all.

_____

Skip Counting to Add and Subtract

Family Note

Today your child found patterns on the number grid and used them to add and subtract.

Please return this Home Link to school tomorrow.

(1) Start at 5. Count by 5s.
Draw an X on each square to show the count.

									0
1	2	3	4	X̶5̶	6	7	8	9	1X̶0̶
11	12	13	14	1X̶5̶	16	17	18	19	2X̶0̶
21	22	23	24	2X̶5̶	26	27	28	29	3X̶0̶
31	32	33	34	3X̶5̶	36	37	38	39	4X̶0̶
41	42	43	44	4X̶5̶	46	47	48	49	5X̶0̶

(2) Use the number grid.
Count by 5s to find the missing numbers.

$5 + 5 =$ _____

$20 - 15 =$ _____

Practice

Solve. Use the number grid above to help you.

(3) Start at 17. Count up 7 hops. Where do you land? _____

(4) Start at 21. Count up 10 hops. Where do you land? _____

(5) Start at 25. Count up 14 hops. Where do you land? _____

(6) Start at 28. Count up 18 hops. Where do you land? _____

Frames-and-Arrows Diagrams

Family Note

Your child is bringing home an activity that may not be familiar to you. It is called "Frames and Arrows." A Frames-and-Arrows diagram shows a sequence of numbers that follow a given rule. Each frame contains a number. The arrow stands for the rule that determines the next number in the sequence or the number that goes in the next frame.

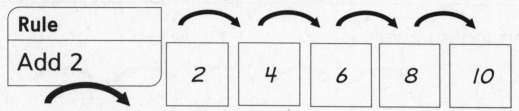

Rule
Add 2

2   4   6   8   10

The rule for the Frames-and-Arrows diagram above is "Add 2," "+ 2," or "Count by 2s."

To solve the Frames-and-Arrows problem below, use the rule to find the missing numbers.

Example 1:

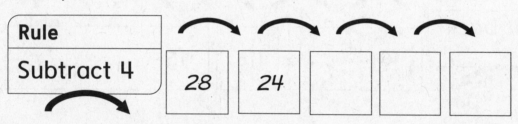

Rule
Subtract 4

28   24

Solution: Write 20, 16, and 12 in the empty frames.

For the next Frames-and-Arrows diagram, look at the number sequence. Determine the arrow rule.

Example 2:

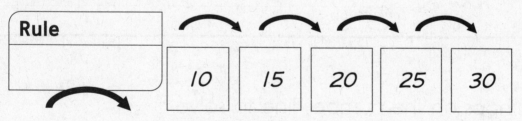

Rule

10   15   20   25   30

Solution: The arrow rule is "Add 5," "+ 5," or "Count by 5s."

Your child has been solving problems like Example 1, in which the arrow rule is given and some of the numbers in the frames are missing. In the next lesson, children will do problems like Example 2, in which the numbers in the frames are given and the arrow rule is missing.

Frames-and-Arrows Diagrams (cont.)

Family Note

Ask your child to tell you about Frames and Arrows. Take turns making up and solving
Frames-and-Arrows problems.

Please return this page of the Home Link to school tomorrow. Save page 77 for future reference.

Find the missing numbers.

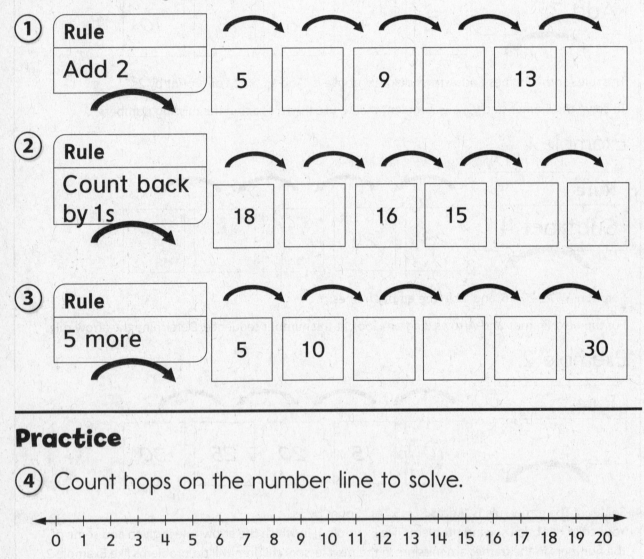

① **Rule** Add 2

| 5 | | 9 | | 13 | |

② **Rule** Count back by 1s

| 18 | | 16 | 15 | | |

③ **Rule** 5 more

| 5 | 10 | | | | 30 |

Practice

④ Count hops on the number line to solve.

0 1 2 3 4 5 6 7 8 9 10 11 12 13 14 15 16 17 18 19 20

$6 + 3 =$ _____   _____ $= 10 - 2$

_____ $= 9 - 5$ $7 - 4 =$ _____

Finding the Rule

Family Note

Today your child worked with Frames-and-Arrows diagrams that had missing rules. You may wish to refer to the Family Note for Lesson 3-9 to review the Frames-and-Arrows routine.

Please return this Home Link to school tomorrow.

Show someone at home how to find the rules.
Then write each rule.

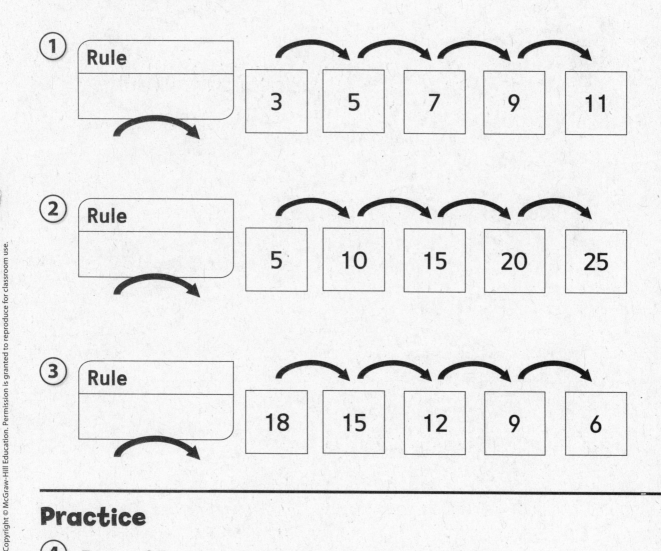

① Rule

3   5   7   9   11

② Rule

5   10   15   20   25

③ Rule

18   15   12   9   6

Practice

④ Draw 35 stars on the back of the page.

Counting with Calculators

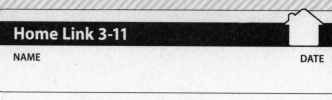
Family Note

Today your child learned how to program a calculator to count up and back by different numbers. Children connected this counting to addition and subtraction. If you have a calculator at home, program it to skip count and use it to check your child's work below.

Please return this Home Link to school tomorrow.

Write the counts.

(1) Start at: 3
Count: up
By: 3s

———— , ———— , ———— , ———— , ———— , ———— , ————

(2) Start at: 20
Count: back
By: 2s

———— , ———— , ———— , ———— , ———— , ———— , ————

(3) Start at: 45
Count: back
By: 5s

———— , ———— , ———— , ———— , ———— , ———— , ————

Practice

(4) Write number sentences to show the domino sums.

————————   ————————

Length and Addition Facts

Two ideas are emphasized in Unit 4: length measurement and addition fact fluency.

Children begin the unit by directly comparing the lengths of two objects. Then they compare the lengths of two objects indirectly by using a third object, such as a piece of string. Later children learn to measure length using nonstandard units like paper clips.

They learn that measurement units must be the same size.

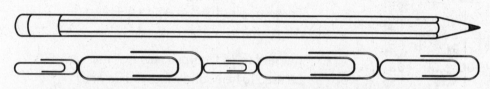

Using different-size units does not provide an accurate measurement.

They also learn that the units must be arranged without gaps or overlaps.

Measuring with gaps and overlaps does not provide an accurate measurement.

Correct measures use same-size units with no gaps and overlaps.

The pencil is about 4 paper clips long.

Also in this unit, children transition from displaying data in tally charts to displaying data in bar graphs. Their work with comparing lengths will help them interpret data by comparing the lengths of the bars in the graphs.

Other lessons in Unit 4 focus on addition facts. One of the Grade 1 standards requires children to fluently add and subtract within 10. In order to achieve fluency, they must be efficient at recalling these facts and using the facts in a variety of situations. Doubles and combinations of 10 are some of the easiest facts for children to remember and are emphasized in Unit 4. Once children learn these facts, they can use them to help figure out other facts. Fact fluency is emphasized and developed throughout the year, so do not worry if your child does not achieve this goal right away.

In Unit 4, children also begin developing strategies for adding more than two numbers and using place value to mentally add or subtract 10 from other 2-digit numbers.

Vocabulary Important terms in Unit 4:

bar graph A graph with bars that represent data.

What time do we eat dinner?

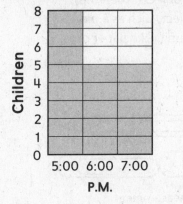

addition facts Two numbers from 0 to 10 and their sum, such as 9 + 7 = 16.

combinations of 10 Addition facts in which the numbers add to 10. For example, 4 + 6 = 10 and 3 + 7 = 10 are combinations of 10.

doubles Addition facts in which both numbers being added are the same. For example, 4 + 4 = 8 and 9 + 9 = 18 are doubles.

helper fact A fact you know well that can be used to help solve a fact you do not know well.

Do-Anytime Activities

To work with your child on concepts taught in this and previous units, try these activities:

1. Measure flat objects in your home using paper clips. For example, you might measure the length of your mobile phone, the width of a small table, or the length of a spoon. Work with your child to place the paper clips end-to-end, without gaps or overlaps.

2. Use your fingers to help your child practice finding combinations of 10. For example, show both hands with 2 fingers up and the rest closed. Your child should tell you that you have 2 fingers up and 8 fingers down. Continue with different finger combinations. You can also practice doubles facts this way by placing a number of fingers up, and asking your child to tell you double that number of fingers.

3. Draw a bar graph like the one shown above, but list three activities your child likes to do after school along the bottom, such as play with friends, ride bikes, and read. Have your child keep track of the number of times he or she does each activity in a given week. For example, if your child comes home and plays with friends, he or she should color up to the number 1 above "play with friends" on the bar graph. At the end of the week, discuss which activity your child did most often and least often.

Building Skills through Games

Below are some of the games your child will play in Unit 4:

Fishing for 10

Each player draws 5 number cards. The object is to "fish for" pairs that add to 10.

Roll and Record Doubles

Each player rolls a die, doubles the number that was rolled, and records the total on a chart. The game ends when one column of the chart is filled.

What's Your Way?

Players take turns mentally finding 10 more and 10 less than a given number and sharing their strategies for doing so.

As You Help Your Child With Homework

As your child brings home assignments, you may want to go over the instructions together, clarifying them as necessary. The answers listed below will guide you through the Home Links for this unit.

Home Link 4-1

1–2. Answers vary.

3. Sample answer: No. Everything in Problem 1 is longer than the string, so the things in Problem 1 are longer than the things in Problem 2.

4. 9

Home Link 4-2

1–4. Answers vary.   **5.** 55

Home Link 4-3

1–3. Answers vary.

4. 13   **5.** 14   **6.** 12   **7.** 16

Home Link 4-4

1–3. Answers vary.   **4.** 12; 5 + 7 = 12

Home Link 4-5

1. Answers vary.   **2.** 7; 3 + 4 = 7

Home Link 4-6

1. 5   **2.** 4   **3.** Before bedtime; 2

4. 11, 13, 15, 17, 19

Home Link 4-7

1–4. Answers vary.

Home Link 4-8

1. Answers vary.   **2.** 6 pennies

Home Link 4-9

1–4. Answers vary.   **5.** 17; 9 + 8 = 17

Home Link 4-10

1–5. Answers vary.   **6.** 11, 10

Home Link 4-11

1. 33   **2.** 13   **3.** 48   **4.** 28

5. 12; 8 + 4 = 12

Introducing Length Measurement

Family Note

Today your child began exploring length. Children directly compared the lengths of two objects. Then they compared the lengths of two objects indirectly by using a third object, such as a strip of paper or a piece of string. Ask your child to explain how to use a strip of paper to compare length.

Please return this Home Link to school tomorrow.

① List 3 things at home that are *longer* than the string.

_____

_____

② List 3 things at home that are *shorter* than the string.

_____

_____

③ Is anything you listed for Problem 1 shorter than anything you listed for Problem 2? Explain.

_____

_____

_____

_____

Practice

④ Draw the missing dots. Fill in the sentence.

_____ + 8 = 17

Measuring Length

Family Note

Today your child learned that length is measured with same-size units placed end to end without gaps or overlaps. As your child completes the Home Link, check that the units are placed end to end without gaps or overlaps.

Please return this Home Link to school tomorrow.

Find at least 5 of the same item. They must be the same size. For example, you might choose paper clips, blocks, or toothpicks. This will be your length unit.

(1) I chose _____ for my length unit.

(2) Draw a line that is 1 _____ long.

(3) Draw a line that is 3 _____ long.

(4) Draw a line that is 5 _____ long.

Practice

(5) Count up to solve.

What number is 12 more than 43? _____

seventy-nine   79

More Length Measurement

Family Note

In the last lesson, your child measured objects by lining up same-size units (such as toothpicks) end to end. Today, children explored ways to measure objects using only one unit, such as an unsharpened pencil. They practiced moving the pencil along the length of an object, avoiding gaps and overlaps, and counting the number of pencils to determine the length.

Please return this Home Link to school tomorrow.

(1) Trace the outline of your thumb in the box.

This is a _____ thumb
length unit.   your name

(2) About how many thumbs
long is this pencil?

About _____ thumbs

(3) About how many thumbs
long is this string?

About _____ thumbs

Practice

Count up to solve.

(4) What number is 6 more than 7? _____

(5) What number is 5 more than 9? _____

(6) What number is 8 more than 4? _____

(7) What number is 7 more than 9? _____

Measuring with Spoons

Use spoons to find the length of each object.

(1) Your bed   About _____ spoons long

(2) Your pillow   About _____ spoons long

(3) A sofa or chair   About _____ spoons long

Draw or write how you used the spoons to measure.

Practice

(4) Ming has 5 limes and 7 oranges.

How many pieces of fruit does Ming have?

_____ pieces of fruit

Number model: _____

Building Shapes from Shapes

① Trace these triangles onto another paper.
Cut them out from that paper.
Combine them to make four different shapes.
Draw all four shapes you made on the back of this paper.

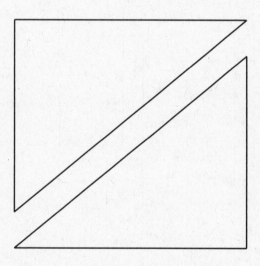

Practice

② Shana has 3 green marbles and 4 red marbles.
How many marbles does Shana have in all?

_____ marbles

Number model: _____

Representing Data with a Bar Graph

Family Note

In previous lessons, children organized data in tally charts. Today your child learned how to use a bar graph to organize and interpret data.

Please return this Home Link to school tomorrow.

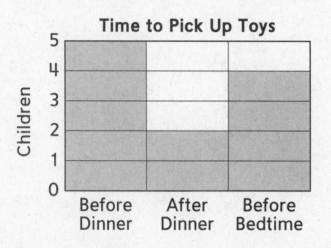

Time to Pick Up Toys

Answer the questions about the bar graph.

1. How many children pick up toys before dinner?
   _____ children

2. How many children pick up toys before bedtime?
   _____ children

3. Did more children pick up toys after dinner
   or before bedtime? _____

   How many more children? _____ more children

Practice

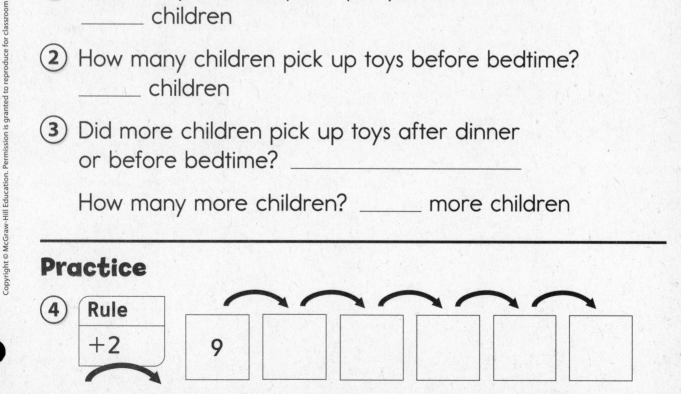

4.

Rule	
+2	

9

Introducing Doubles

Family Note

Today your child was introduced to doubles such as 1 + 1 = 2, 2 + 2 = 4, 3 + 3 = 6, and so on. Doubles are often some of the first addition facts that children master.

Please return this Home Link to school tomorrow.

Play *Roll and Record Doubles* with someone at home.

Directions

① Roll a dot die. Use that number to make a double.

② Shade the first empty box above the sum for the double.

③ Take turns until one column is filled.

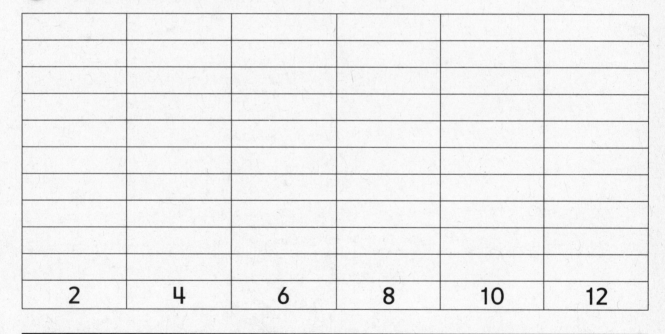

2	4	6	8	10	12

Practice

④ Draw dots on the domino.

Write a number sentence for the domino.

_____

Family Letter

Addition and Subtraction Facts in *First Grade Everyday Mathematics*

In Unit 4, children are formally introduced to addition facts, defined as two numbers from 0 to 10 and their sums, such as $9 + 7 = 16$. Subtraction problems using the same numbers, such as $16 - 7 = 9$ and $16 - 9 = 7$, are known as subtraction facts, which will be formally introduced later in first grade. Learning addition and subtraction facts is a major focus of first grade mathematics. Future work with addition and subtraction builds on these basic facts, and many strategies children develop for solving their basic facts can later transfer to computation with larger numbers. *Everyday Mathematics* supports children's progress toward fluency with addition and subtraction facts by encouraging children to do the following:

- Put numbers together and take them apart flexibly, for example, by seeing that 8 is the same as $6 + 2$, $4 + 4$, $3 + 5$, and so on.

- Discover and compare efficient strategies for solving basic facts.

- Practice basic facts in meaningful ways, through number stories, Quick Looks with ten frames, and games.

Knowing doubles ($2 + 2$, $3 + 3$, $4 + 4$, and so on) and combinations of 10 ($1 + 9$, $2 + 8$, $3 + 7$, and so on) can help children solve nearly all other addition or subtraction facts. For this reason, these two groups of facts are a major focus in *First Grade Everyday Mathematics*. In Units 6 and 7, children learn strategies for solving more difficult facts.

As your child solves basic fact problems or plays fact games at home, you may wish to support his or her development of fact fluency by asking questions, such as these:

- How did you figure it out?

- Can you say aloud how you thought about it in your head?

- Is there another way you could figure it out?

- If someone did not know the answer, how would you explain to that person how to figure it out?

Discussion and practice with good fact strategies in first grade will lead to eventual mastery of all basic facts.

Combinations of 10

Family Note

Earlier in the year, your child explored pairs of numbers that add to 10. Recognizing these pairs of numbers helps with mastering many addition facts. Today your child worked with these number pairs again and categorized them formally as a group of facts known as *combinations of 10*.

Please return this Home Link to school tomorrow.

① Do Two-Fisted Penny Addition with someone at home:

- Make a pile of 10 pennies.

- The person at home grabs some pennies with one hand and keeps that hand closed.

- Count the pennies left in the pile and decide how many pennies are in the closed hand.

- Fill in the table. Record more on the back of this sheet.

Number of Pennies in the Pile	Number of Pennies in the Closed Hand	Number Model
8	2	8 + 2 = 10

Practice

② How many pennies? Remember to include a unit.

_____

More Combinations of 10

Family Note

Your child spent more time today working with combinations of 10. Continue working with your child to help him or her recognize combinations of 10.

Please return this Home Link to school tomorrow.

Do this activity with someone at home.

1. Someone at home says a number between 0 and 10.

2. You say the number that makes a combination of 10.

3. Then say the addition fact and record the number sentence in the space below.

4. Repeat at least 6 more times.

Practice

5. A fish tank has 9 goldfish and 8 rainbow fish.

    How many fish are in the tank? _____ fish

    Write a number model.

    Number model: _____

Adding Three Numbers

Family Note

Today your child learned about adding three numbers. Children learned that when more than two numbers are added, it does not matter which pair of numbers is added first.

Please return this Home Link to school tomorrow.

Solve. Then answer the questions.

$4 + 2 + 8 =$ _____

① Which numbers did you add first? _____

② Why did you add these first?

_____

$3 + 7 + 7 =$ _____

③ Which numbers did you add first? _____

④ Why did you add these first?

_____

⑤ What other way could you add them?

_____

_____

Practice

⑥ Solve.

```
  <--|----|----|----|----|----|----|-->
     5    6    7    8    9   10   11
```

Start at 7. Count up 4 hops. Where do you land? _____

Start at 5. Count up 5 hops. Where do you land? _____

10 More, 10 Less

Family Note

Today your child began exploring ways to find 10 more or 10 less using a number grid. By the end of the year, children should be able to find 10 more or 10 less than a number in their heads. They will continue to practice throughout the year.

Please return this Home Link to school tomorrow.

1	2	3	4	5	6	7	8	9	10
11	12	13	14	15	16	17	18	19	20
21	22	23	24	25	26	27	28	29	30
31	32	33	34	35	36	37	38	39	40
41	42	43	44	45	46	47	48	49	50

Solve.

(1) What number is 10 more than 23? _____

(2) What number is 10 less than 23? _____

(3) What number is 10 more than 38? _____

(4) What number is 10 less than 38? _____

Practice

(5) April has 8 crayons.
José has 4 crayons.
How many crayons do April and José have in all?
Write a number model.

_____ crayons

Number model: _____

Place Value and Comparisons

In Unit 5, children begin to use larger numbers and explore place value. They learn that the digits in a 2-digit number represent the number of tens and ones. For example, in the number 72, 7 is in the tens place and has a value of 7 tens, or 70; 2 is in the ones place and has a value of 2 ones, or 2. Children use base-10 blocks to represent numbers and to demonstrate their understanding of place value by exchanging 10 ones for 1 ten, and vice versa.

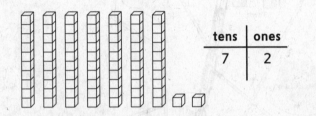

tens	ones
7	2

Children compare numbers using the symbols $<$, $>$, and $=$. They discuss what the equal sign ($=$) means and how to use it. They determine whether number sentences are true or false.

These number sentences are true:	These number sentences are not true; they are false:
$2 + 9 = 9 + 2$	$4 + 5 = 3 + 7$
$4 + 7 = 15 - 4$	$9 - 8 = 1 + 1$
$3 + 3 = 1 + 5$	$13 - 4 = 10 - 9$
$10 = 10$	$7 = 12 - 8$

Children go from comparing numbers with $<$ and $>$ to modeling comparison number stories. In comparison stories, they decide which of two quantities is larger and then find the difference. Children use situation diagrams to help make sense of these problems.

Children also spend more time on measurement. They measure different sections of a crooked path and add the lengths to find the total length of the path.

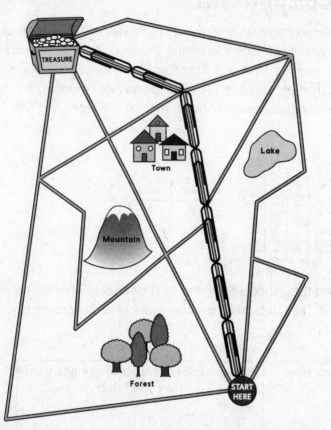

Please keep this Family Letter for reference as your child works through Unit 5.

Vocabulary   Important terms in Unit 5:

comparison diagram A diagram used in *Everyday Mathematics* to model situations in which two quantities are compared. The diagram contains two quantities and their difference.

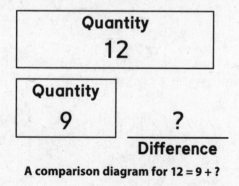

A comparison diagram for 12 = 9 + ?

cube In *First Grade Everyday Mathematics,* a base-10 block that represents 1.

long In *First Grade Everyday Mathematics,* a base-10 block that represents 10.

number scroll A series of number grids taped together.

Do-Anytime Activities

To work with your child on the concepts taught in this unit and in previous units, try these activities:

1. Look for 2-digit numbers in and around your home. Ask your child to tell you how much each digit is worth. Ask your child to compare 2-digit numbers using <, >, and =.

2. Tell addition and subtraction number stories using 1- and 2-digit numbers of household objects. Then work together to solve them and write number models. Discuss strategies.

3. Ask your child to order a group of items in your home from shortest to longest.

Building Skills through Games

In Unit 5, your child will play these and other games to develop skills with addition, place value, and comparing numbers:

Addition Top-It

In this *Top-It* variation, each player turns over and adds two cards. The higher sum wins the round.

Base-10 Exchange

Players take turns putting base-10 blocks on their Tens-and-Ones Mat according to the roll of a die. Whenever possible, they exchange 10 cubes for 1 long. The first player to get 10 longs wins.

The Difference Game

Players each pick a card and collect as many pennies as the number shown on the card. Then players count each other's pennies and figure out how many more pennies one player has than the other.

The Digit Game

Each partner draws two cards from a deck of number cards. The player whose cards make the larger 2-digit number takes all of the cards drawn. The player with more cards at the end of the game wins.

Stop and Go

There is a GO player and a STOP player. The GO player tries to *go* to 50, adding 1- and 2-digit numbers. The STOP player tries to *stop* the GO player by subtracting 10 and 20 from 2-digit numbers.

Top-It with Relation Symbols

In this *Top-It* variation, children compare their cards using <, >, and = cards.

As You Help Your Child with Homework

As your child brings home assignments, you may want to go over the instructions together, clarifying them as necessary. The answers listed below will guide you through the Home Links for this unit.

Home Link 5-1
1. 56   **2.** 40   **3.** 12   **4.** Answers vary.

Home Link 5-2
1. Sample answer: 61, 62, 63, 64, 65

2. Sample answer: 48, 18, 28, 78, 68

3. 8; $4 + 4 = 8$

Home Link 5-3
1. 1 Ⓓ and 0 Ⓟ; 10

2. 1 Ⓓ and 2 Ⓟ; 12

3. Sample answer: 2 Ⓓ and 1 Ⓟ; 21

4. Answers vary.

Home Link 5-4
1. >; <; =; <; >; <

2. 17; $8 + 9 = 17$

Home Link 5-5
1. False; True; True; True; False; True; False

2. ③1; ⑨4; ①7

Home Link 5-6
1–2. Answers vary.

3.

									100
101	102	103	104	105	106	107	108	109	110
111	112	113	114	115	116	117	118	119	120
121	122	123	124	125	126	127	128	129	130

4. 6; 7; 6

Home Link 5-7
1. Answers vary.

2. 63   **3.** 19   **4.** 72

Home Link 5-8
1.

| | .
or
• • • • • • • • • •

2. Answers vary.

Home Link 5-9
1. >; >; <; =; <; =; <; >

2. 9; $3 + 2 + 4 = 9$

Home Link 5-10
1. Bart; 4; Sample number model: $12 - 8 = 4$

2. Martha; 7; Sample number model: $3 + 7 = 10$

3. Answers vary.

Home Link 5-11
1. 40; $60 - 20 = 40$

2. 85; $54 + 31 = 85$

3. 70; $56 + 14 = 70$

4. False

Home Link 5-12
Sample answers given for problems 1–6.

1. Hammer   **2.** Scissors   **3.** Computer

4. Number line   **5.** Counters   **6.** Coins

7. 4; 6; 9

Introducing Place Value

Family Note

Today your child learned about place value using base-10 blocks. In the charts below, the blocks in the Tens box are called *longs,* and the blocks in the Ones box are called *cubes.* Ten cubes is the same as one long. Base-10 blocks are used throughout *Everyday Mathematics* to represent multidigit numbers.

Please return this Home Link to school tomorrow.

Example:

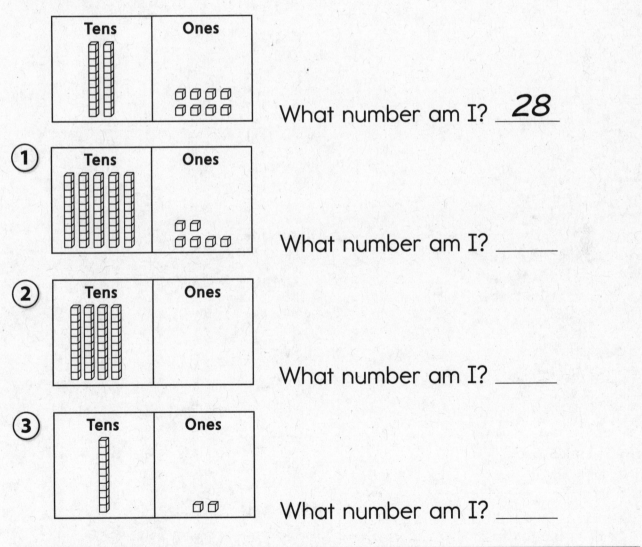

Tens	Ones

What number am I? __28__

(1)

Tens	Ones

What number am I? _____

(2)

Tens	Ones

What number am I? _____

(3)

Tens	Ones

What number am I? _____

Practice

(4) Use a pencil to measure a large box.
How tall is the box? About _____ pencils

Digits and Place Value

Family Note

Today your child explored place value using calculators and number grids. Children used a calculator to see how digits change as we count, specifically when we count from 9 to 10, 39 to 40, and so on. Then children used a number grid to observe the relationship between numbers that have the same digit in the tens place or the same digit in the ones place.

IMPORTANT: Please send at least 5 dimes to class with your child tomorrow. Your child will continue exploring place value using pennies and dimes tomorrow.

Please return this Home Link to school tomorrow.

① List 5 numbers with 6 in the tens place.

② List 5 numbers with 8 in the ones place.

Practice

③ Oliver and Olivia each have 4 rings.

How many rings do they have in all?

_____ rings

Number model: _____

Pennies, Dimes, and Place Value

1 cent	10 cents
Ⓟ	Ⓓ

① Ⓟ Ⓟ Ⓟ Ⓟ Ⓟ Ⓟ Ⓟ Ⓟ Ⓟ Ⓟ is the same as _____ Ⓓ and _____ Ⓟ.

This is _____ cents.

② Ⓟ Ⓟ Ⓟ Ⓟ Ⓟ Ⓟ Ⓟ Ⓟ Ⓟ Ⓟ Ⓟ Ⓟ is the same as _____ Ⓓ and _____ Ⓟ.

This is _____ cents.

③ Ⓟ Ⓟ Ⓟ Ⓟ Ⓟ Ⓟ Ⓟ Ⓟ Ⓟ Ⓟ Ⓟ Ⓟ Ⓟ Ⓟ Ⓟ Ⓟ Ⓟ Ⓟ Ⓟ is the same as _____ Ⓓ and _____ Ⓟ.

This is _____ cents.

Practice

④ How many spoons are in your kitchen? _____ spoons

Relation Symbols

Family Note

Today your child was introduced to the relation symbols < and >. The < means "is less than," and the > means "is more than." These symbols are used in the same way = is used to mean "is equal to" or "is the same amount as." For example, instead of writing 5 *is less than* 8, we write 5 < 8.

It takes time for children to learn how to correctly use these symbols. One way to help your child identify the correct symbol is to draw two dots near the larger number and one dot near the smaller number. Then connect the dots as shown below.

Another way is to think of the open end of the symbol as a mouth eating the larger number.

Please return this Home Link to school tomorrow.

---

① Write <, >, or =.

Example: 18 ____>____ 12

> < is less than
> > is more than
> = is the same amount as
> = is equal to

11 _____ 7   21 _____ 25   37 _____ 37

29 _____ 42   35 _____ 15   48 _____ 78

Practice

② Talia has 8 red leaves.
Jon has 9 yellow leaves.
How many leaves do they have in all? _____ leaves

Number model: _____

one hundred eleven   111

The Equal Sign

Family Note

Today your child continued practicing addition and subtraction and working with the equal sign as he or she determined whether number sentences were true or false. Your child also changed numbers and symbols (+, −, =, <, >) to make number sentences true.

Please return this Home Link to school tomorrow.

(1) Write *True* or *False* next to each number sentence.

$10 = 7 + 2$ _____

$4 + 4 = 3 + 5$ _____

$10 - 5 = 0 + 5$ _____

$3 + 9 = 9 + 3$ _____

$14 - 7 = 8$ _____

$7 = 7$ _____

$4 + 0 = 3 - 1$ _____

Practice

(2) Circle the tens digit in each number.

3 1

9 4

1 7

Number Scrolls

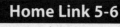

Family Note

Today your child used knowledge of place value to fill in number grids and then construct number scrolls. Ordering numbers on a grid helps children identify number patterns and develop number sense. Talk with your child about patterns in the number grid shown below.

Please return this Home Link to school tomorrow.

① Tell someone at home how you filled in number grids to make a number scroll.

② Ask about any other kinds of scrolls that person knows.

③ Show that person how to fill in the bottom 3 rows of this number grid.

									100
101									
				115					
									130

Practice

④ Solve.

$2 + 4 =$ _____

_____ $= 10 - 3$

$4 +$ _____ $= 10$

Measuring
Crooked Paths

Family Note

Today your child learned to measure the length of a crooked path. Children found that the length of a path is the same whether they measure the whole path at once or measure each of its parts and add the lengths together. This understanding will help children measure more complex paths.

Please return this Home Link to school tomorrow.

(1) Use one paper clip to measure the length of this path. Write a number model to show adding the parts of the path.

This path is _____ paper clips long.

Number model: _____

Practice

What numbers do the base-10 blocks show?

(2) _____

(3) _____

(4) _____

Explorations and Exchanges

Family Note

Today your child learned a game involving exchanges with base-10 blocks and explored comparing and measuring length. Have your child tell you about the Explorations that the class did today.

Please return this Home Link to school tomorrow.

(1) This is one way to show the number 21 with base-10 blocks.

I............

Use I and ▪ to show 21 in two other ways.

Practice

(2) Use a fork to measure.

How many forks wide is your kitchen sink?

_____ forks

More Comparison Symbols

Family Note

Today your child practiced using relation symbols <, >, and = to model number stories about the weights of various animals.

Please return this Home Link to school tomorrow.

(1) Fill in the blank with <, >, or =.

12 _____ 11

13 + 20 _____ 31

28 _____ 19 + 10

15 _____ 9 + 6

7 _____ 17

45 _____ 45

17 + 3 _____ 22

40 _____ 20 + 0

Practice

(2) Sandra's cat had 3 gray kittens, 2 spotted kittens, and 4 white kittens.

How many kittens did she have in all? _____ kittens

Number model: _____ + _____ + _____ = _____

one hundred twenty-one   121

Comparison Number Stories

Family Note

Today your child used comparison diagrams to model comparison number stories and find the difference between two numbers. Just as with other number story situations, comparison diagrams are provided to help children organize their thinking as they begin to rely less on real objects.

For example:

Mary has 2 pennies. Pablo has 5 pennies.
Who has more pennies? How many more?

Pablo has 3 more pennies than Mary.

Please return this Home Link to school tomorrow.

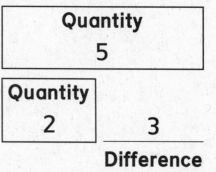

Solve. Use the diagrams to help you.
Then write a number model to match.

① Bart has 12 pennies. Perry has 8 pennies.

Who has more pennies? _____

How many more? _____ pennies

Number model: _____

> Quantity
>
> Quantity 8
>
> _____
> **Difference**

② Tricia has 3 pennies. Martha has 10 pennies.

Who has more pennies? _____

How many more? _____ pennies

Number model: _____

> Quantity
>
> Quantity 3
>
> _____
> **Difference**

Practice

③ How many pillows are in your home? _____ pillows

Two-Digit Addition and Subtraction

Family Note

Today your child solved addition and subtraction number stories about animal weights. For the problems below, encourage your child to explain different methods he or she could use to solve the number stories (such as using a number line or number grid).

Please return this Home Link to school tomorrow.

Add or subtract to solve the animal number stories.

(1) How much taller is a peacock (60 in.) than an owl (20 in.)?

$$||||||| - || = ||||$$

_____ in.   Number model: _____ − _____ = _____

(2) How long would the sun bear (54 in.) and parrot (31 in.) be if they lay nose to nose?

$$||||| \ \bullet\bullet\bullet\bullet + ||| \ \bullet = |||||||| \ \bullet\bullet\bullet\bullet\bullet$$

_____ in.   Number model: _____ + _____ = _____

(3) How much do a beaver (56 lb) and a fox (14 lb) weigh all together?

$$||||| \ \bullet\bullet\bullet\bullet\bullet\bullet + | \ \bullet\bullet\bullet\bullet = ||||||||$$

_____ lb   Number model: _____ + _____ = _____

Practice

(4) True or False? $7 > 4 + 3$ _____

Using Tools

Ask someone at home to tell you about three tools they use at home or at work. Write the tools here.

① _____

② _____

③ _____

Write three tools that you use in math class.

④ _____

⑤ _____

⑥ _____

Tell someone at home how you use one of the tools.

Practice

⑦ Solve.

$13 - \underline{\quad} = 9$

$14 - \underline{\quad} = 8$

$16 - \underline{\quad} = 7$

Addition Fact Strategies

In Unit 6, children continue to work with addition facts and develop strategies for solving more difficult facts. For example, many children quickly learn the doubles addition facts: $1 + 1 = 2$; $2 + 2 = 4$; $3 + 3 = 6$; and so on. Using doubles facts, they learn to solve nearby facts using the *near doubles* strategy. A child who knows $4 + 4$ can use it to solve $5 + 4$ by thinking of it as a double plus 1, or $3 + 4$ by thinking of it as a double minus 1. These "helper facts" are a useful tool for solving other addition facts.

Children also gain experience with an important strategy for mentally adding numbers. *Making 10* is a strategy that involves breaking apart one addend, making a ten, and then adding what remains to 10. For example, children learn to add $8 + 6$ by breaking apart the 6: $8 + \mathbf{6} = 8 + \mathbf{2} + \mathbf{4} = 10 + 4 = 14$. This strategy takes advantage of properties of addition that can help children add more efficiently.

Also in Unit 6, children apply their skills with number stories and place value to continue building strategies for solving 2-digit addition problems.

Children also begin telling time to the hour on analog clocks. Digital clocks and time to the half hour will be introduced in the next unit.

Please keep this Family Letter for reference as your child works through Unit 6.

Vocabulary   Important terms in Unit 6:

flat In *Everyday Mathematics,* a base-10 block that represents 100.

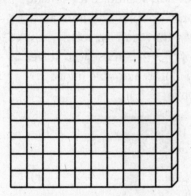

making 10 A method or strategy of mentally adding two numbers by breaking apart one addend to make ten, then adding what remains to 10. For example,
$7 + 4 = 7 + 3 + 1 = 10 + 1 = 11.$

name-collection box In *Everyday Mathematics,* a diagram that is used for collecting equivalent names for numbers.

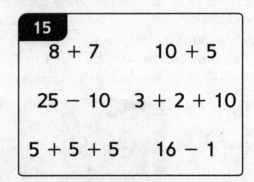

near doubles An addition strategy that involves using a known doubles fact to solve a nearby fact. For example, $5 + 4 = 9$ is *near* the doubles $4 + 4 = 8$ and $5 + 5 = 10$, so either double could be used to find the sum of $5 + 4$.

Do-Anytime Activities

To work with your child on the concepts taught in this unit and in previous units, try these activities:

1. Have your child tell number stories that fit given equations, such as $8 + 5 = 13$ and $7 + 7 = 14$.

2. Fill in name-collection boxes. Begin with a number, such as 20, and have your child provide at least five equivalent names.

3. Encourage your child to show you how to use the "making 10" strategy to solve $7 + 5$. Have him or her suggest other facts that could be solved using this strategy.

4. Ask your child to tell time to the hour using analog clocks.

Building Skills through Games

Your child will play these games and others in Unit 6:

Fishing for 10

Each player draws 5 number cards. The object is to "fish" for pairs of numbers that add to 10.

Penny-Dime-Dollar Exchange

Players roll two dice and put that number of cents on their Place-Value Mats. Whenever possible, they exchange 10 pennies for 1 dime. The first player to make an exchange for a $1 bill wins.

Roll and Record Doubles

Players roll a die and make a double with that number. The first player to fill a column on the record sheet wins.

As You Help Your Child with Homework

As your child brings home assignments, you may want to go over the instructions together, clarifying them as necessary. The answers listed below will guide you through the Home Links for this unit.

Home Link 6-1
5. 14 stickers; $7 + 4 + 3 = 14$

Home Link 6-2
1. Answers vary.
2. Jordan's pencil

Home Link 6-3
1–2. Answers vary.
3. Sample answer: My shapes have different numbers of sides.
4. 40; 38; 55

Home Link 6-4
1.

Fact	Helper Fact	Answer
Example: $5 + 6 = ?$	$5 + 5 = 10$ *or* $6 + 6 = 12$	$5 + 6 = 11$
$3 + 4 = ?$	Sample answer: $3 + 3 = 6$	$3 + 4 = 7$
$5 + 4 = ?$	Sample answer: $5 + 5 = 10$	$5 + 4 = 9$
$7 + 8 = ?$	Sample answer: $7 + 7 = 14$	$7 + 8 = 15$

2. $3 = 3$; $4 = 9 - 5$; $10 + 2 = 12$

Home Link 6-5

1. 8

2. 9

3. Sample answer: I know 4 + 4 = 8, so 1 more is 9.

4. 6; 2; 9

Home Link 6-6

1. Check your child's picture to make sure the answers are correct and it is colored correctly.

2. 10; 9; 2

Home Link 6-7

1. Answers vary.

2. 9; 7; 8

Home Link 6-8

1. 3

2. 14

3. 5

4. <; =; >

Home Link 6-9

1. 0 + 10; 10 + 0; 1 + 9; 9 + 1; 2 + 8; 8 + 2; 3 + 7; 7 + 3; 4 + 6; 6 + 4; 5 + 5

2. Sample answers: 20 − 5; 5 + 5 + 5; 17 − 2; 6 + 9

3. <; >; <; =

Home Link 6-10

1. 92

2. 48

3. 9

4. 8 > 18; 15 = 5 + 6; 11 − 3 = 14

Home Link 6-11

1. Sample answer: $1 $1 D D D D D D P P P P P

2. Sample answer: $1 $1 $1 D D P P P P

3. 111¢; $1.11

4. 17 balls, 8 + 6 + 3 = 17

The Hour Hand

Family Note

Today your child observed how the hour hand on an analog clock moves as an hour passes. For now, children focus on telling time to the hour while looking at the hour hand only. Later in the year, they will be introduced to the minute hand.

Please return this Home Link to school tomorrow.

Draw the hour hand.

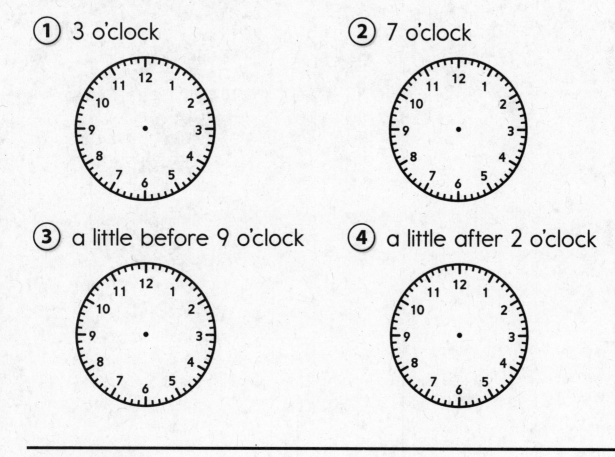

(1) 3 o'clock

(2) 7 o'clock

(3) a little before 9 o'clock

(4) a little after 2 o'clock

Practice

(5) Bao has 7 dog stickers, 4 cat stickers, and 3 dinosaur stickers.

How many stickers does Bao have in all? _____

Number model: _____ + _____ + _____ = _____

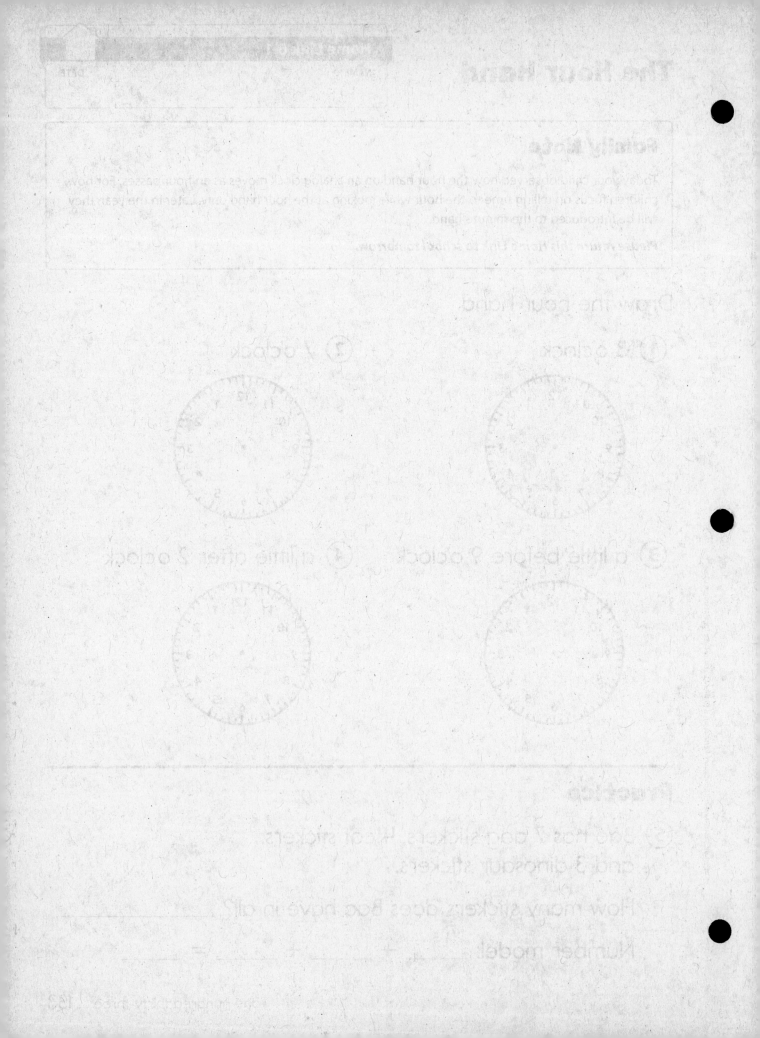

Number Stories

Example:

I have 4 balloons.
Jamal brought 1 more balloon.
We have 5 balloons together.

Number model: 4 + 1 = 5

Unit
balloons

① Find a picture from a magazine, or draw your own picture. Use it to write a number story. Write a number model to go with your story.

Unit

Number model: _____

Practice

② Alex's pencil is longer than Brianna's pencil.
Alex's pencil is shorter than Jordan's pencil.
Whose pencil is the longest?

_____

Shapes Made from Shapes

Family Note

Today your child determined whether number sentences are true or false, practiced addition doubles, and created shapes with given attributes. They will continue this work more formally in future units.

Please return this Home Link to school tomorrow.

① Draw a shape using 2 triangles.
Draw a different shape using 2 triangles.

② Draw a shape using a rectangle and a triangle.
Draw another shape using a rectangle and a triangle.

③ How do you know you made a different shape?

_____

_____

Practice

④ Find 10 more than each of the numbers.
Think about counting up by 10s to help you.

30 _____   28 _____   45 _____

Near Doubles

Family Note

Today your child learned how to use a doubles fact, such as $8 + 8 = 16$, to help solve facts close to doubles, such as $8 + 9 = 17$. This strategy is called *near doubles*. Children learned about helper facts (facts that are useful for solving other facts) in Unit 4 and will continue to learn about other helper facts as the year progresses.

Please return this Home Link to school tomorrow.

(1) Write a helper fact and then the final answer for each number sentence in the table below.

Fact	Helper Fact	Answer
Example: $5 + 6 = ?$	$5 + 5 = 10$ *or* $6 + 6 = 12$	$5 + 6 = 11$
$3 + 4 = ?$		
$5 + 4 = ?$		
$7 + 8 = ?$		

Practice

(2) Circle the **true** number sentences.

$3 = 3$ $4 = 9 - 5$ $6 = 3 + 2$

$4 = 7 - 2$ $10 + 2 = 12$ $4 + 9 = 12$

one hundred thirty-nine   139

Recording Near-Doubles Strategies

Family Note

Today your child spent more time using doubles facts to help solve nearby facts called *near doubles*. Children focused on explaining their solution strategies using words, pictures, or number sentences. Ask your child to explain how he or she solved the number stories below either with words or with a written number sentence.

Please return this Home Link to school tomorrow.

Solve the number stories.

① Tommy had 4 pretzels.
His mom gave him 4 more pretzels.
How many pretzels does Tommy have now?

_____ pretzels

② Renee had 4 pretzels.
Her mom gave her 5 more pretzels.
How many pretzels does Renee have now?

_____ pretzels

③ How can you use the first number story to help you solve the second number story?

Practice

④ Complete each number sentence.

$3 + 3 =$ _____ $8 +$ _____ $= 10$ $10 = 1 +$ _____

Finding Addition Sums

Family Note

Today, your child continued to explore strategies for solving addition facts. Children practiced the making-10 strategy. Ask your child to explain how the making-10 strategy works and how to find each sum on this Home Link.

Please return this Home Link to school tomorrow.

① Solve. Use the color code to color the picture.

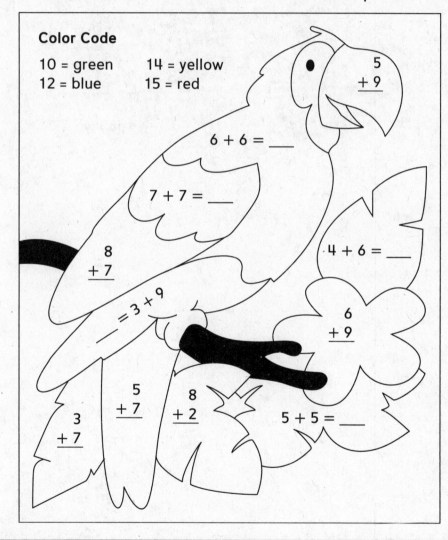

Color Code
10 = green   14 = yellow
12 = blue   15 = red

$6 + 6 =$ ___

$7 + 7 =$ ___

$\begin{array}{r} 8 \\ + 7 \\ \hline \end{array}$

$\begin{array}{r} 5 \\ + 9 \\ \hline \end{array}$

$4 + 6 =$ ___

___ $= 3 + 9$

$\begin{array}{r} 6 \\ + 9 \\ \hline \end{array}$

$\begin{array}{r} 3 \\ + 7 \\ \hline \end{array}$

$\begin{array}{r} 5 \\ + 7 \\ \hline \end{array}$

$\begin{array}{r} 8 \\ + 2 \\ \hline \end{array}$

$5 + 5 =$ ___

Practice

② Complete the following number sentences.

$3 + 7 =$ _____ $1 +$ _____ $= 10$ _____ $+ 8 = 10$

one hundred forty-three   143

Comparison Number Stories

Family Note

Today your child explored *My Reference Book,* an important *Everyday Mathematics* resource that children can use to find out more about the mathematics they learn in class. Ask your child to show you how to use the table of contents so you can explore *My Reference Book* together.

Please return this Home Link to school tomorrow.

① Write or draw a comparison number story.

Then solve.

Example:

Yasmin bought 6 stickers this week.
She bought 8 stickers last week.
How many more stickers did Yasmin buy
last week than this week?

$8 - 6 = 2$ more stickers

Practice

② Solve.

$18 -$ _____ $= 9$ $7 +$ _____ $= 14$ _____ $= 4 + 4$

Number Stories

Family Note

Today your child learned that when solving a problem, it is helpful to think about what the problem is asking. This is called *making sense of a problem*. Making sense of a problem can help your child decide what needs to be done to solve the problem. For example, should the numbers in the problem be added or subtracted?

Throughout the year, encourage your child to explain how he or she knew what to do when solving a problem.

Please return this Home Link to school tomorrow.

Solve.

(1) You have 12 marbles.
Your friend has 15 marbles.
How many more marbles does your friend have?
_____ marbles

(2) You picked 6 red flowers and 8 blue flowers.
How many flowers did you pick in all? _____ flowers

(3) 7 children were at the playground.
2 children went home.
How many children stayed at the playground?
_____ children

Tell someone at home how you knew what to do to solve the stories.

_____

Practice

(4) Use <, >, or = to make each number sentence true.

13 _____ 27   44 _____ 44   80 _____ 30

one hundred forty-seven   147

Name-Collection Boxes

Family Note

Today your child began working with name-collection boxes. See the attached Family Letter for more information about this *Everyday Mathematics* routine.

Please return this Home Link to school tomorrow.

(1) List all of the addition facts you know that have a sum of 10.

MRB
53

_____

_____

(2) Write as many names as you can in the name-collection box.

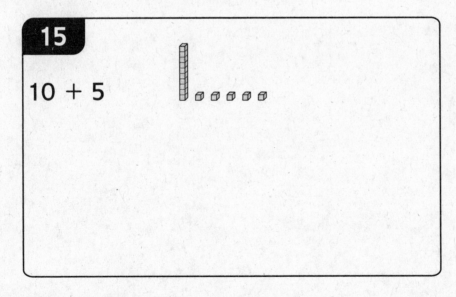

15

10 + 5

Practice

(3) Write <, >, or =.

57 _____ 81   95 _____ 65

30 _____ 50   77 _____ 77

Name-Collection Boxes

People, things, and ideas often have several different names. For example, Mary calls her parents Mom and Dad. Other people may call them Linda and John, Aunt Linda and Uncle John, or Grandma and Grandpa. Mail may come addressed to Mr. and Mrs. West. All of these names are for the same two people.

Numbers can also be called by many names. Your child is bringing home an activity with a special format for recording different number names. We call this format a **name-collection box.** The box is used by children to collect many names for a given number.

The box is identified by the name on the label. The box shown below is a 25-box, or a name-collection box for the number 25.

A name-collection box can be filled using any equivalent names. In first grade, children focus mostly on sums (20 + 5), differences (35 − 10), and combinations of operations (10 + 10 + 10 − 5). Children check whether these names are correct by writing number sentences and asking themselves whether the number sentences are true or false. For example, 20 + 5 is a name for 25 because 20 + 5 = 25 is a true number sentence. Alternatively, 40 − 10 is not a name for 25 because 40 − 10 = 25 is a false number sentence. Names can also include words (even words in other languages), tally marks, and combinations of coins.

25
37 − 12   20 + 5
⫶⫶⫶⫶ ⫶⫶⫶⫶ ⫶⫶⫶⫶ ⫶⫶⫶⫶ ⫶⫶⫶⫶
twenty-five veinticinco
10 + 10 + 10 − 5

More Base-10 Riddles

Solve the riddles.

MRB
71-72

Example:

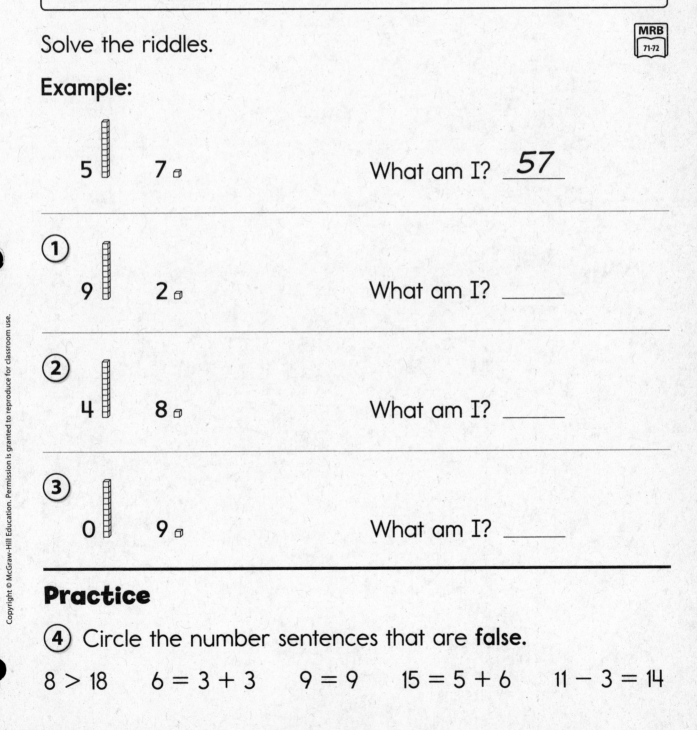

5   7   What am I? __57__

① 9   2   What am I? _____

② 4   8   What am I? _____

③ 0   9   What am I? _____

Practice

④ Circle the number sentences that are **false**.

8 > 18   6 = 3 + 3   9 = 9   15 = 5 + 6   11 − 3 = 14

Dollars and Place Value

Family Note

Today your child used money to think about place value. Children exchanged ones, tens, and hundreds using pennies, dimes, and dollars. Because children have just begun to work with dollars, some of the problems on this page may be difficult for your child. If possible, use real money to model the problems. Start by counting the bills and coins in the example with your child.

Please return this Home Link to school tomorrow.

MRB
110-111

Show how you would pay for each item.
Use $1, D, or P.

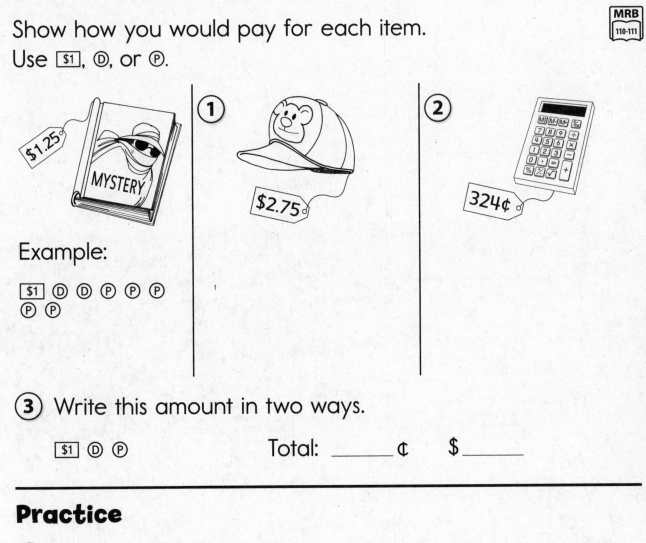

Example:

$1 D D P P P
P P

(3) Write this amount in two ways.

$1 D P   Total: _____ ¢ $ _____

Practice

(4) Kyle bought 8 tennis balls, 6 baseballs, and 3 golf balls. How many balls did he buy in all?

_____   _____ + _____ + _____ = _____

one hundred fifty-five   155

Subtraction Fact Strategies and Attributes of Shapes

In Unit 7, children continue to work on addition and subtraction fact fluency. They begin by looking at fact families, which show related addition and subtraction facts. Children practice facts with Fact Triangles and through games. They also explore and compare strategies for subtracting. They solve "What's My Rule?" problems that require adding and subtracting to complete number patterns.

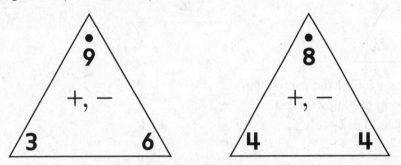

Fact Triangles for 3, 6, 9, and 4, 4, 8

Also in this unit, children explore attributes of shapes and determine which attributes define shapes. For example, a square can be blue, red, or green and still be a square. Therefore, color is a *nondefining attribute* for a square. However, a shape is not a square unless it has 4 equal sides. So, having 4 equal sides is a *defining attribute* of a square.

Finally, children connect times shown on digital clocks with those on analog clocks. They examine how the minute hand moves as an hour passes. In this unit, children continue to tell time to the nearest hour on an analog clock (using hour and minute hands) and on a digital clock. In Unit 8, they will start telling time to the half hour.

Please keep this Family Letter for reference as your child works through Unit 7.

Vocabulary   Important terms in Unit 7:

attribute A feature of an object or common feature of a set of objects. Examples of attributes include size, shape, color, and number of sides.

digital clock A clock that shows the time with numbers of hours and minutes, usually separated by a colon.

fact family A set of related arithmetic facts involving the same numbers. *For example:*

$$3 + 4 = 7 \qquad 4 + 3 = 7$$
$$7 - 3 = 4 \qquad 7 - 4 = 3$$

Fact Triangle In *First Grade Everyday Mathematics,* a triangular flash card labeled with the numbers of a fact family that children can use to practice addition/subtraction facts. The addends and their sum appear in the corners of each triangle.

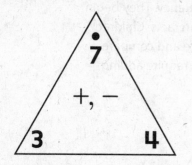

think addition A strategy for solving subtraction facts that involves thinking of a known addition fact. For example, to solve $14 - 7 = \underline{\hspace{1cm}}$, children might think of $7 + 7 = 14$.

"What's My Rule?" A problem involving inputs, outputs, and a rule. Two of the three are known, and the third can be found out.

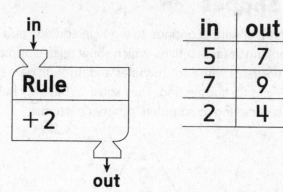

Do-anytime Activities

To work with your child on the concepts taught in this unit and in previous units, try these activities:

1. Use the Fact Triangles that children will begin receiving in Lesson 7-2 to help your child practice addition and subtraction facts.

2. Look for shapes around the house, at the supermarket, in architectural features, and on street signs. Name these shapes using their geometric names, and have children share defining attributes of the shapes.

3. Draw a name-collection box with a number on the tag. Ask your child to write at least 10 addition and subtraction names for the given number.

12
$17 - 5$
$2 + 10$
$4 + 8$
$13 - 1$
twelve
doce
‖‖‖ ‖‖‖ ‖

Building Skills through Games

Your child will play these games and others in Unit 7:

Attribute Train

One player chooses a block to start a sequence, like cars of a train. Players alternate adding a block that differs by only one attribute—shape, thickness, size, or color—from the previous block.

Beat the Calculator

A "Calculator" (a player who uses a calculator) and a "Brain" (a player who does not use a calculator) race to see who will be first to solve addition problems.

Salute!

Without looking at their cards, two players each hold a number card, ranging from 0 to 10, to their foreheads with the number facing out. A third player announces the sum of the two numbers. The first player with a number card to name the number on his or her forehead (without looking) wins the round.

Tric-Trac

Players take turns rolling dice and covering equivalent sums on the gameboard.

As You Help Your Child with Homework

As your child brings home assignments, you may want to go over the instructions together, clarifying them as necessary. The answers listed below will guide you through the Home Links for this unit.

Home Link 7-1

1. 7, 5, 12

$12 - 7 = 5$

$12 - 5 = 7$

$7 + 5 = 12$

$5 + 7 = 12$

2. 6, 9, 15

$15 - 6 = 9$

$15 - 9 = 6$

$6 + 9 = 15$

$9 + 6 = 15$

3. Answers vary.

Home Link 7-3

Sample number models given for Problems 1-4.

1. 3; $3 +$ _____ $= 6$

2. 3; $7 +$ _____ $= 10$

3. 6; $12 = 6 +$ _____

4. 9; $1 +$ _____ $= 10$

5. Answers vary.

Home Link 7-4

1. 5, 8, 13

$5 + 8 = 13$

$8 + 5 = 13$

$13 - 5 = 8$

$13 - 8 = 5$

2. Sample answers: $10 + 4$; $7 + 7$; $15 - 1$

3.
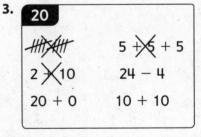

4. 40; 92; 39

Home Link 7-5

1. Sample answers: 6 sides, 6 angles, white color

2. Sample answers: 4 sides, the top and bottom are the same length, gray color

3. 12; 9

Home Link 7-6

1–3. Answers vary.

4. 53

Home Link 7-7

1. Answers vary.

2. Sample answers: color, size, which ways they point

3. Sample answers: 3 sides, all sides are straight, 3 vertices, closed

4. 7; 7; 18

Home Link 7-8

1. + 1; 20; 10

2. − 2; 10; 9

3. − 7; 5; 4

4. 4

5. 4

6. 7

Home Link 7-9

1. + 1

2. − 4

3. + 6

4. − 5

5. 9 hops; 19 − 9 = 10

Home Link 7-10

1. 14; 6; 7; 10

2. 13; 6; 15; 10

3. 50

Home Link 7-11

1. 3:00

2. 11:00

3.

4.

5. <; >; =

Fact Families

Family Note

Today your child generated addition and subtraction facts from dominoes to create fact families. Fact families show related facts and help children relate addition to subtraction. Although most dominoes have two addition facts and two subtraction facts, children discussed fact families for doubles (for example, $4 + 4 = 8$), which have only one addition fact and one subtraction fact.

Please return this Home Link to school tomorrow.

MRB
46

Write the 3 numbers for each domino.
Use the numbers to write a fact family.

(1) Numbers: _____, _____, _____

Fact Family:

_____ − _____ = _____   _____ + _____ = _____

_____ − _____ = _____   _____ + _____ = _____

(2) Numbers: _____, _____, _____

Fact Family:

_____ − _____ = _____   _____ + _____ = _____

_____ − _____ = _____   _____ + _____ = _____

Practice

(3) Use a paper clip. Measure the lengths of two shoes.

My shoe: _____ paper clips

Someone else's shoe: _____ paper clips

Whose shoe is longer? How much longer?

_____ shoe is _____ paper clips longer.

Fact Triangles

This Family Letter includes several pages of Fact Triangles. Each Fact Triangle includes three numbers that make up a fact family. Have your child cut out each Fact Triangle. Use these triangles like flash cards to practice addition and subtraction facts.

The number below the dot is the sum of the other two numbers. For example, 8 is the sum of 5 and 3.

You can help your child practice addition by covering the sum. Your child then adds the numbers that are not covered. For example, if you cover 8, your child adds 5 and 3 to find the sum, 8.

By covering one of the numbers at the bottom of the triangle, your child can practice subtracting the two uncovered numbers on the triangle from their sum. For example, if you cover 3, your child subtracts 5 from 8. If you cover 5, your child subtracts 3 from 8.

Covering one of the numbers at the bottom of the triangle can also be used to practice finding missing addends. For example, if you cover 3, your child determines the number that is added to 5 to get 8. In other words, $5 + \square = 8$.

Fact Triangles have two advantages over regular flash cards:

1. They reinforce the link between addition and subtraction.

2. They help simplify memorization by linking four facts together. Knowing a single fact means you know four facts.

   $5 + 3 = 8$

   $3 + 5 = 8$

   $8 - 5 = 3$

   $8 - 3 = 5$

Save the Fact Triangles in an envelope or a plastic bag, and use them to continue practicing addition and subtraction facts with your child when you have time.

Fact Triangles 1

Cut out the 6 triangles. Practice the addition and subtraction facts on these triangles with someone at home.

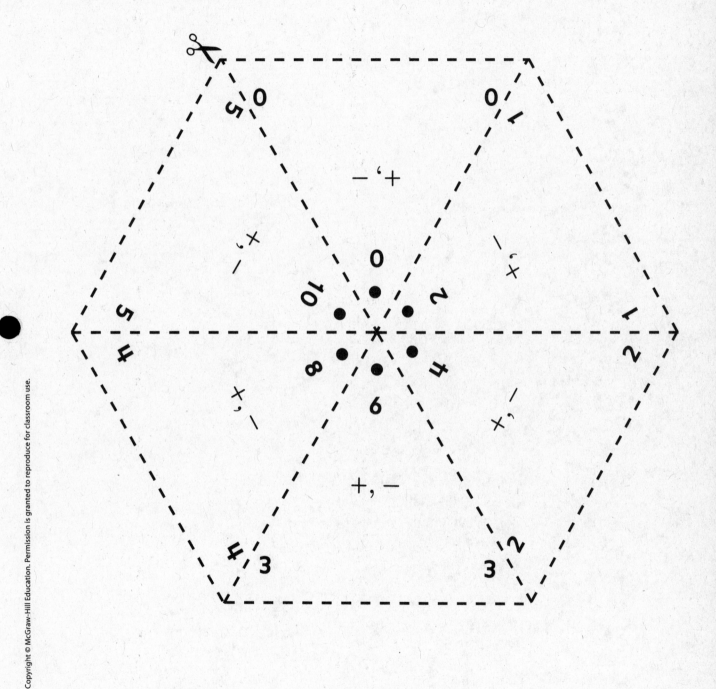

one hundred sixty-five   165

Fact Triangles 2

Cut out the 6 triangles. Practice the addition and subtraction facts on these triangles with someone at home.

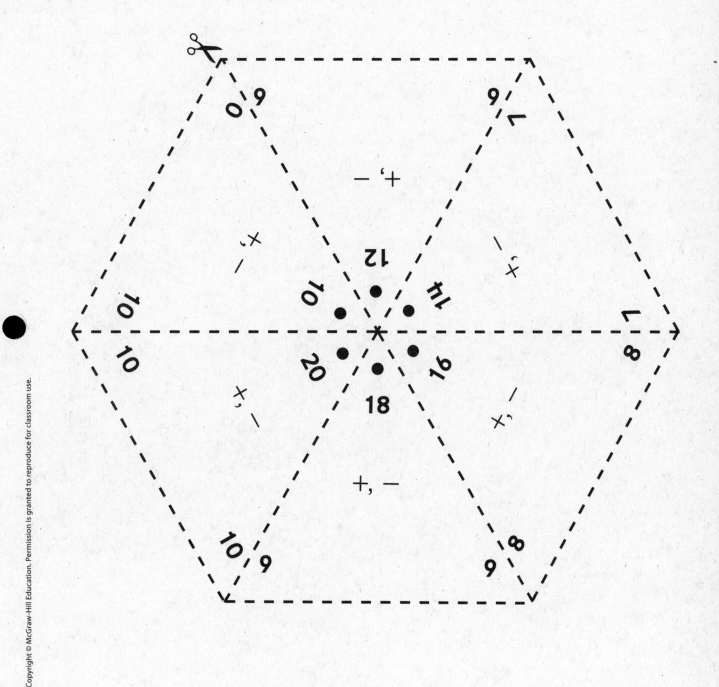

Fact Triangles 3

Cut out the 6 triangles. Practice the addition and subtraction facts on these triangles with someone at home.

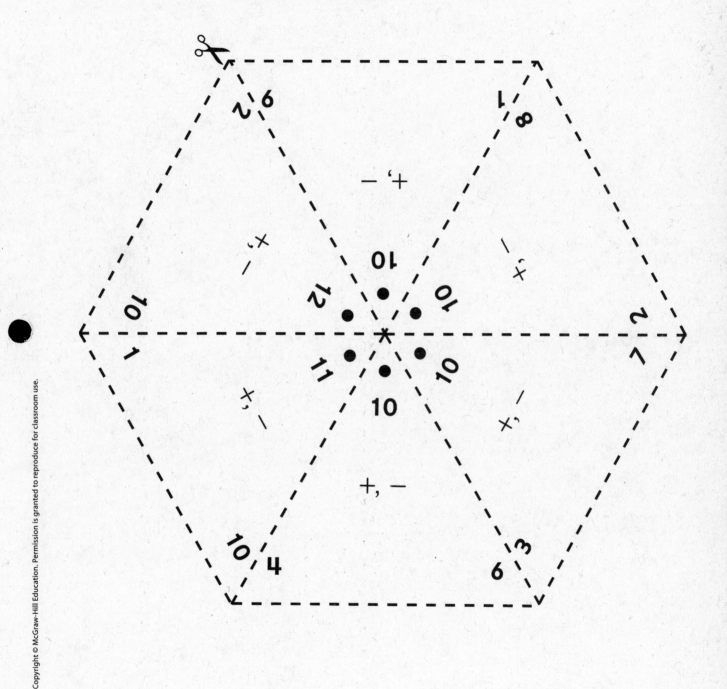

Fact Triangles 4

Cut out the 6 triangles. Practice the addition and subtraction facts on these triangles with someone at home.

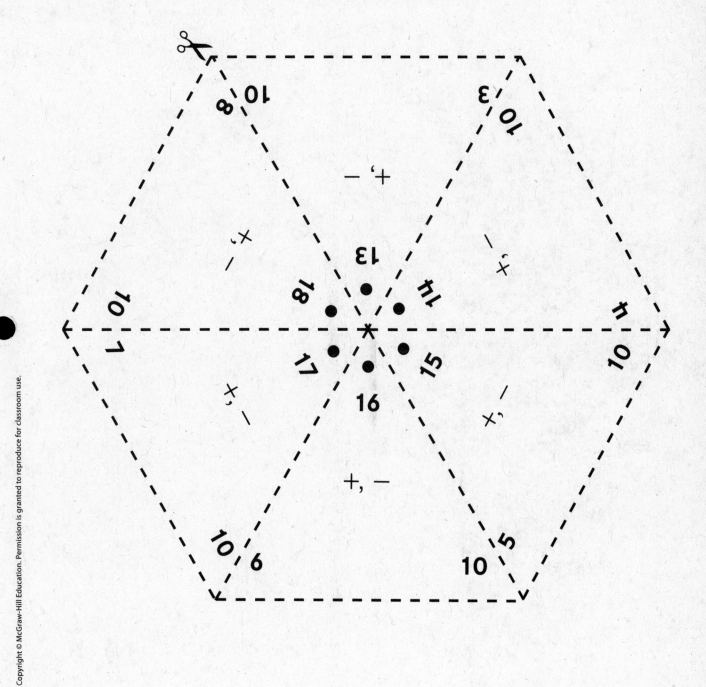

Relating Special Addition and Subtraction Facts

Family Note

In recent lessons, children learned about fact families and how addition facts have related subtraction facts. In today's lesson, your child solved subtraction problems by thinking about related addition facts, particularly with doubles and combinations of 10. For example, children might solve $18 - 9 = \square$ by thinking addition: $9 + \square = 18$.

Please return this Home Link to school tomorrow.

Write an addition fact you can use to find the answer. Then write the answer in the blank.

MRB
48

Example: $16 - 8 = \underline{\quad 8 \quad}$

$8 + \underline{\quad\quad} = 16$

① $6 - 3 = \underline{\quad\quad}$

_____

② $10 - 7 = \underline{\quad\quad}$

_____

③ $12 - 6 = \underline{\quad\quad}$

_____

④ $10 - 1 = \underline{\quad\quad}$

_____

Practice

⑤ Record the time you do each activity.

_____ : _____   _____ : _____   _____ : _____

    wake up   eat lunch   go to bed

More Subtraction Fact Strategies

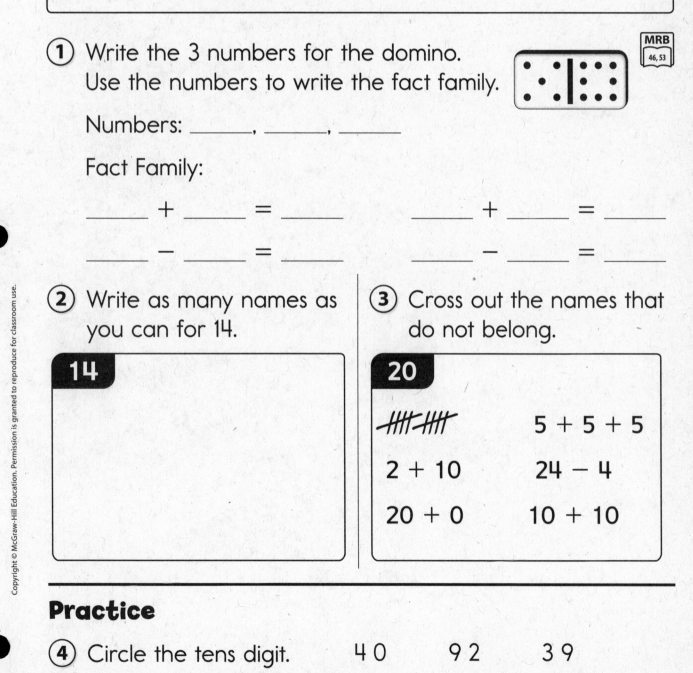

① Write the 3 numbers for the domino.
Use the numbers to write the fact family.

MRB
46, 53

Numbers: _____, _____, _____

Fact Family:

_____ + _____ = _____   _____ + _____ = _____

_____ − _____ = _____   _____ − _____ = _____

② Write as many names as you can for 14.

14

③ Cross out the names that do not belong.

20	
~~⊬⊬⊬ ⊬⊬⊬~~	5 + 5 + 5
2 + 10	24 − 4
20 + 0	10 + 10

Practice

④ Circle the tens digit.   4 0   9 2   3 9

Fact Triangles 5

Cut out the triangles for addition and subtraction fact practice.

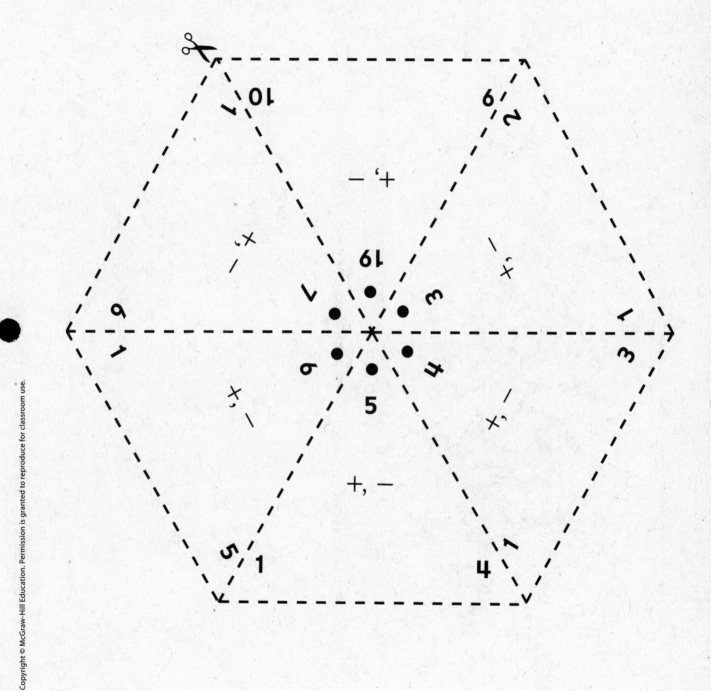

Fact Triangles 6

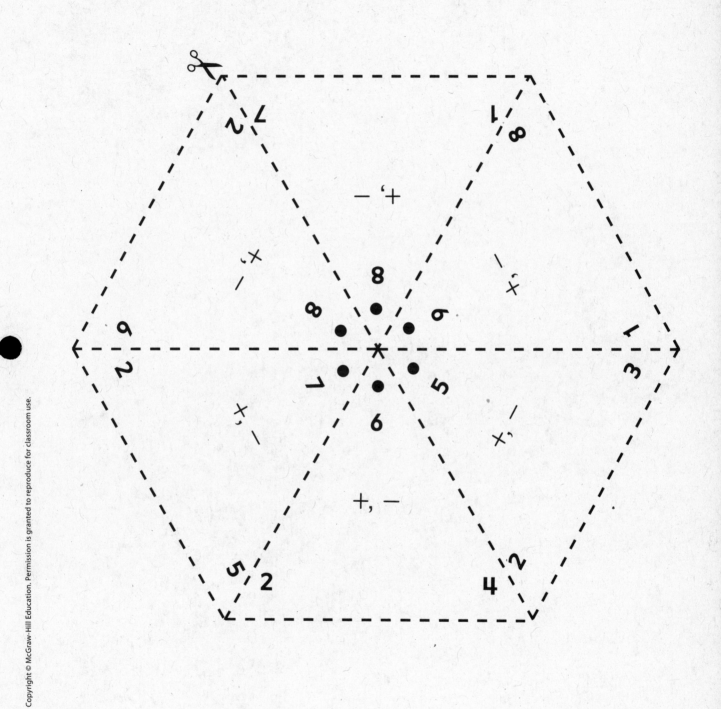

Attributes
of Shapes

Family Note

Today your child explored attributes of shapes. Some attributes of shapes are color, size, or number of sides or corners. Encourage your child to look carefully at objects all around—not just geometric objects—and identify their attributes.

Please return this Home Link to school tomorrow.

List three attributes of each shape.

MRB
125

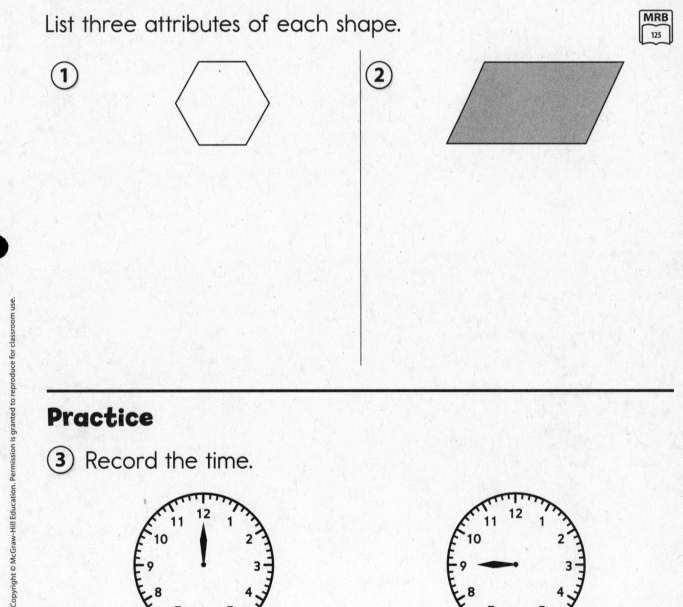

① ②

Practice

③ Record the time.

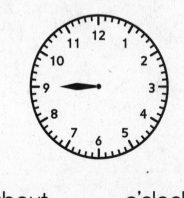

about _____ o'clock   about _____ o'clock

one hundred eighty-one   181

Exploring Attributes, Fractions, and *Salute!*

Family Note

Today your child explored the connection between addition and subtraction in the game *Salute!*, divided shapes in half, and further explored attributes of shapes. Children will continue working with shapes in future lessons. To prepare for this, help your child find objects with the shapes listed below.

Also included in this Home Link are more Fact Triangles for further fact practice.

Please return this Home Link to school tomorrow.

① Find something in your house that has a triangle on it.  Write its name, or draw its picture.

② Find something in your house that has a circle on it. Write its name, or draw its picture.

③ Find something in your house that has a square on it. Write its name, or draw its picture.

Practice

④ What number do the base-10 blocks show?

_____

Fact Triangles 7

Cut out the triangles for addition and subtraction fact practice.

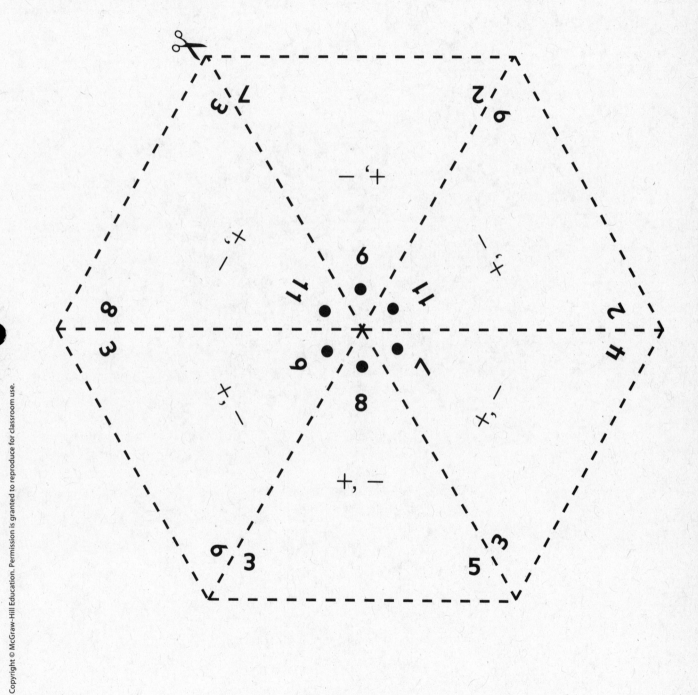

Fact Triangles 8

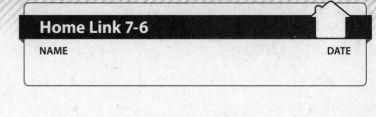

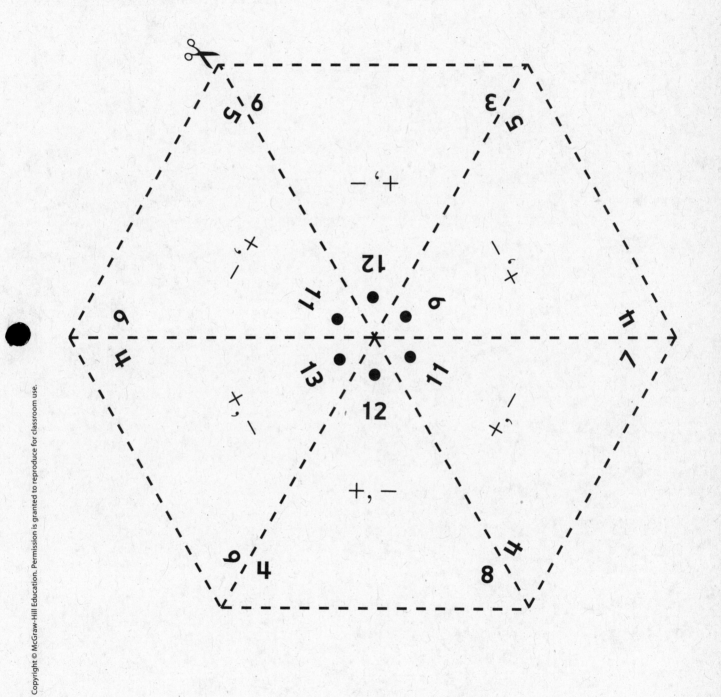

Defining and Nondefining Attributes

Family Note

Today your child learned about the attributes that define triangles and rectangles, such as the numbers of sides and corners (also called *vertices*). Without these defining attributes, a shape cannot be a triangle or a rectangle. Children also learned about nondefining attributes of shapes, such as color and size.

Please return this Home Link to school tomorrow.

① Draw 2 different triangles. The triangles must have *at least two* attributes that are different.

② Name the attributes that are different.

③ Name the attributes that are the same.

Practice

④ Solve.

$2 + 5 =$ _____   _____ $= 5 + 2$ $3 + 7 + 8 =$ _____

"What's My Rule?"

Family Note

Ask your child to explain what the function machine is doing to the "in" numbers before he or she fills in the missing "out" numbers. For example, in the first problem, the function machine is adding 1 to each of the "in" numbers.

Also included in this Home Link are more Fact Triangles. This set of Fact Triangles includes three blanks. Fill them with whatever facts your child would like to practice more.

Please return this Home Link to school tomorrow.

Fill in the missing rule and numbers.

MRB
56-58

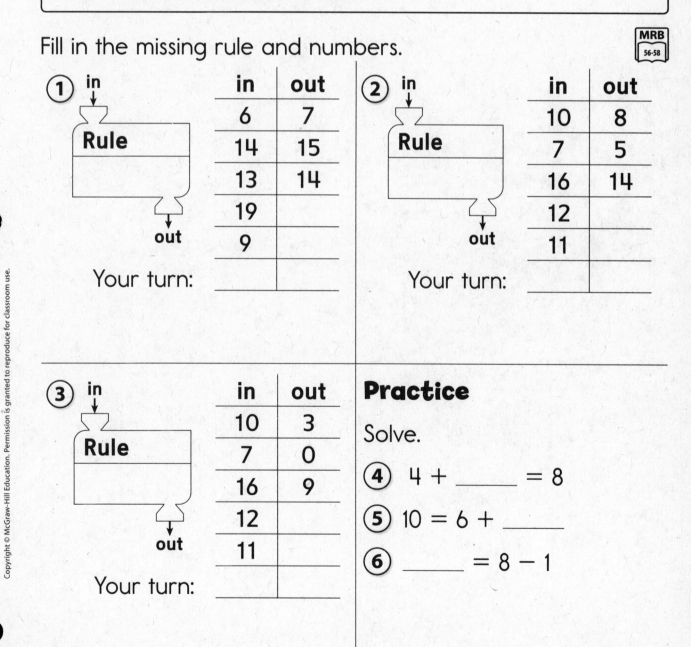

① in
Rule
out
Your turn:

in	out
6	7
14	15
13	14
19	
9	

② in
Rule
out
Your turn:

in	out
10	8
7	5
16	14
12	
11	

③ in
Rule
out
Your turn:

in	out
10	3
7	0
16	9
12	
11	

Practice

Solve.

④ 4 + _____ = 8

⑤ 10 = 6 + _____

⑥ _____ = 8 − 1

Family Letter

"What's My Rule?"

Today your child learned about a kind of problem you may not have seen before. We call it "What's My Rule?" Please ask your child to explain it to you. Here is a little background information you might find useful.

Imagine a machine that has a funnel at the top and a tube at the bottom—we call this a *function machine*. The function machine can be programmed so that when you drop a number into the funnel at the top, the machine changes the number according to the rule and a new number comes out of the tube at the bottom.

For example, you can program the machine to add 2 to any number that is dropped into the funnel. If you put in 3, out comes 5; if you put in 6, out comes 8.

You can show this with a table:

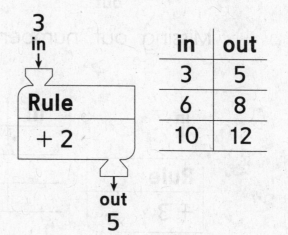

in	out
3	5
6	8
10	12

Here is another example of a function machine:

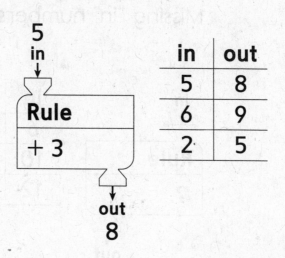

in	out
5	8
6	9
2	5

one hundred ninety-three   **193**

Family Letter

(continued)

In a "What's My Rule?" problem, some of the information is missing. To solve the problem, you have to find the missing information. The missing information can be the numbers that are dropped in, the numbers that come out, or the rule for programming the machine. *For example:*

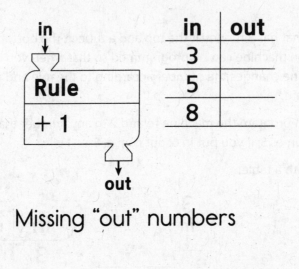

in	out
3	
5	
8	

Missing "out" numbers

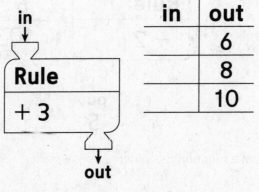

in	out
	6
	8
	10

Missing "in" numbers

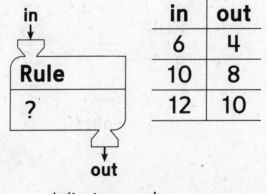

in	out
6	4
10	8
12	10

Missing rule

Fact Triangles 9

Cut out the triangles to use for addition and subtraction fact practice.

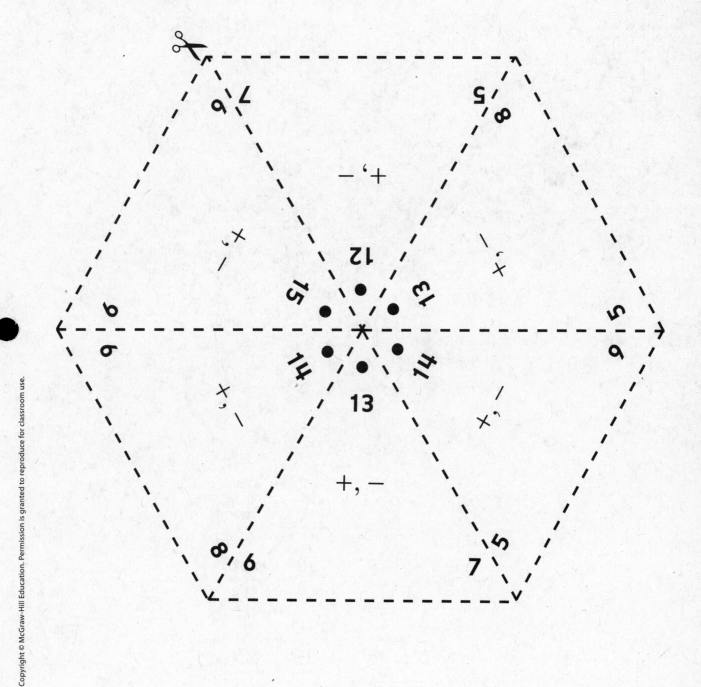

Fact Triangles 10

6

7

9

7

+, −

×, ÷

×, ÷

16

15

17

8

8

×, ÷

×, ÷

+, −

"What's My Rule?"

Find the rules.

① in → Rule → out

in	out
2	3
7	8
3	4
11	12

② in → Rule → out

in	out
9	5
14	10
7	3
4	0

③ in → Rule → out

in	out
1	7
4	10
11	17
8	14

④ in → Rule → out

in	out
15	10
30	25
12	7
9	4

Practice

⑤ Cyrus started at 19 on his number line.
He hopped backward and landed on 10.
How many hops did Cyrus make? _____
Number model: _____

Addition Facts: "What's My Rule?"

Family Note

In previous lessons, children solved "What's My Rule?" problems in which they were asked to find outputs and rules. Today they solved problems in which they had to find inputs. Have your child share strategies for finding the input numbers in Problem 1 below.

Also included in this Home Link are more Fact Triangles. This last set of Fact Triangles are all blanks. Fill them with whatever facts your child would like to practice more.

Please return this Home Link to school tomorrow.

Solve the "What's My Rule?" problems.
Complete the number sentences to check your answers.

MRB 56-58

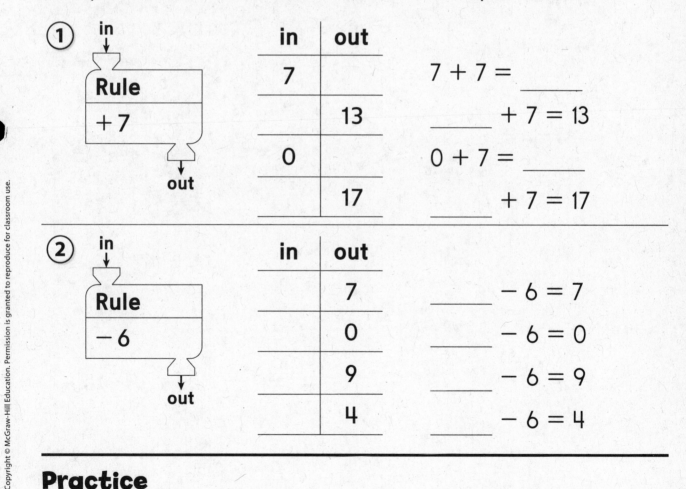

1

Rule
+ 7

in	out
7	
	13
0	
	17

7 + 7 = _____

_____ + 7 = 13

0 + 7 = _____

_____ + 7 = 17

2

Rule
− 6

in	out
	7
	0
	9
	4

_____ − 6 = 7

_____ − 6 = 0

_____ − 6 = 9

_____ − 6 = 4

Practice

3

[4 longs] + [1 long] = _____

Blank Fact Triangles

Cut out the triangles for addition and subtraction fact practice. Fill them with whatever facts you need to practice more.

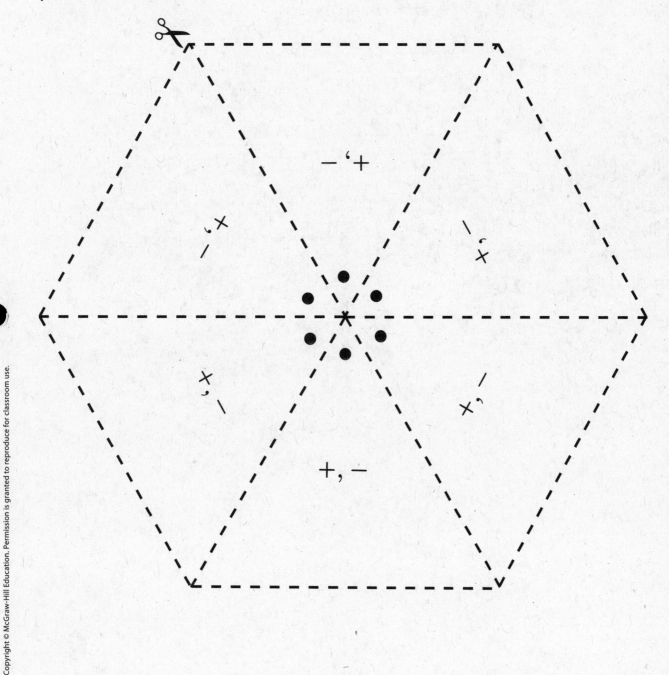

Time on a Digital Clock

Family Note

In Unit 6, your child learned about the hour hand of a clock and how it moves as hours pass. Children told time on clocks that had only hour hands. In today's lesson, your child learned about the minute hand. Children told time to the hour on analog clocks with hour hands and minute hands. They also learned to read the time on digital clocks.

Please return this Home Link to school tomorrow.

Record the time.

MRB
105-106

①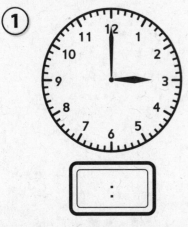

:

②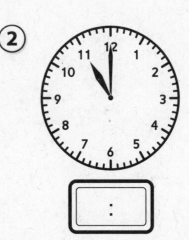

:

Draw hands to show the time.

③

2:00

④

7:00

Practice

⑤ Write <, >, or =.

13 ☐ 42   106 ☐ 105   16 + 23 ☐ 39

two hundred five   205

Geometry

In Unit 7, children began to look carefully at attributes of 2-dimensional shapes. In Unit 8, they extend this work to include 3-dimensional shapes. They also explore building 2- and 3-dimensional shapes. They begin by building shapes with specific attributes, for example, shapes with 4 sides or shapes with 3 corners. Then they learn how to build larger shapes from smaller shapes. This is called *composing shapes*.

In Unit 8, children also learn how to make and name fractions of shapes. Children explore ways to divide shapes into 2 and 4 equal shares. They look at how these shares relate to the whole, and they name each share with a fraction name, including 1 half, 1 out of 2 parts, 1 fourth, 1 quarter, and 1 out of 4 parts. Children also name the whole, using language such as whole, 2 out of 2 parts, 2 halves, 4 out of 4 parts, 4 quarters, and 4 fourths. Children then build on their fraction work, applying

their knowledge of fractions to telling time to the half hour. At this point, children will not be taught the notation typically used with fractions ($\frac{1}{2}$, $\frac{1}{4}$, $\frac{2}{4}$, and so on). This notation will be introduced in Grade 2.

Also in this unit, children continue using place value to add and subtract numbers, including adding and subtracting 10 mentally.

IMPORTANT: Please send a few everyday objects, such as paper towel tubes, balls, books, dice, party hats, or plastic perfume bottles, to school with your child to use as examples for learning about 3-dimensional shapes. Your child will explore these shapes throughout Unit 8.

Please keep this Family Letter for reference as your child works through Unit 8.

Vocabulary   Important terms in Unit 8:

edge A side where two faces meet.

equal shares Another name for equal parts. The result of dividing something into parts that are all the same size.

face A flat surface on a 3-dimensional figure.

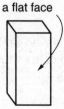

a flat face

fourth When a whole is divided into four equal shares, one-fourth is one of those shares. Also called a *quarter, 1 fourth,* or *1 out of 4 equal shares.*

half When a whole is divided into two equal shares, one-half is one of those shares. Also called *1 half* or *1 out of 2 equal shares.*

half past Thirty minutes after a specific hour. For example, 6:30 is "half past six."

number-grid puzzle In *Everyday Mathematics,* a piece of a number grid in which some of the numbers are missing. Children use number-grid puzzles to practice place-value concepts.

whole An entire object or collection of objects.

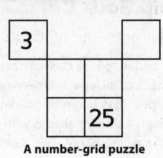

A number-grid puzzle

Do-Anytime Activities

To work with your child on the concepts taught in this unit and in previous units, try these activities:

1. Continue to work on addition and subtraction facts using the Fact Triangles introduced in Unit 7 and games from *My Reference Book.*

2. Encourage your child to build with blocks. Talk about how the pieces fit together to form new shapes and patterns.

3. Have your child tell you the time to the hour and half hour.

Building Skills through Games

Your child will play these games and others in Unit 8:

I Spy

One player describes a shape by naming its attributes. For example, "I spy something with 4 sides." The player continues naming attributes until someone guesses the shape.

Make My Design

Two players start with the same pattern blocks. Player 1 makes a design that Player 2 cannot see. Player 1 describes it to Player 2, who then tries to make the design. Then they check whether the designs are the same. Players switch roles and play again.

Time Match

A player turns over two cards with pictures of analog or digital clocks on them. If the times are not the same, the cards are turned back over. If the cards show the same time, the player keeps the cards. The player with the most cards wins.

As You Help Your Child with Homework

As your child brings home assignments, you may want to go over the instructions together, clarifying them as necessary. The answers listed below will guide you through the Home Links for this unit.

Home Link 8-1
1. Answers vary.

2. <; =; >

Home Link 8-2
1. Sample answers:

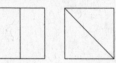

2. Sample answers: half; 1 out of 2 equal shares; 1 half

3. 13; 67

Home Link 8-3
1. Sample answers:

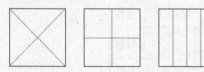

2. Sample answers: quarter; fourth; one out of four equal shares; 1 fourth; 1 quarter

3. 5

Home Link 8-4
1. Sample answers: half, 1 out of 2 parts, one-half, 1 half

2. Sample answers: quarter, fourth, 1 out of 4 parts, one-fourth, one-quarter, 1 fourth

3. 1 out of 2 equal parts

4. 7

5. 8

Home Link 8-5
1. Sample answers:

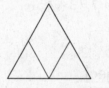

2. Answers vary.

Home Link 8-6
1., 3., 5. Answers vary.

2. 6

4. Square

6. The answers to Problems 2 and 4

7. Sample answer:

Home Link 8-7

1. Sample answer: I know $6 + 6 = 12$, so I add 1 more to get $6 + 7 = 13$.

2. Sample answer: I can take 4 away from 7 and add that to the 6 to make 10. I have 3 left, so $10 + 3 = 13$. 13 is the answer.

3. Answers vary.

4. 7; $12 - 7 = 5$

Home Link 8-8

1. 5

2. 7

3. 2:30

4. 9:30

5. Answers vary.

6. Answers vary.

7. 7; Sample number model: $7 = 3 + 4$

Home Link 8-9

1.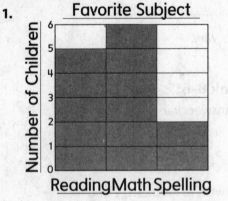
Favorite Subject

2. 13 children

3. 4 children

4. Sample answer:

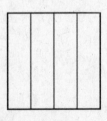

Home Link 8-10

1.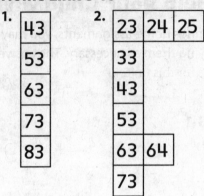

43
53
63
73
83

2.

23	24	25
33		
43		
53		
63	64	
73		

3.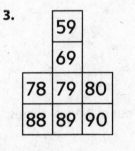

	59	
	69	
78	79	80
88	89	90

4.

14	15	16	17	18
24				28
34		36		38
44				48
54	55	56	57	58

5.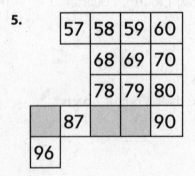

57	58	59	60
	68	69	70
	78	79	80
	87		90
96			

6. 40

Home Link 8-11

1–2. Answers vary.

3. 14; $4 + 2 + 8 = 14$

Building Shapes

Family Note

In today's lesson, your child used straws and twist ties to make polygons with certain attributes. Children also played the game *I Spy* in which the Spy names defining attributes of a shape he or she sees and the other children guess the shape.

Please return this Home Link to school tomorrow.

① Play at least 4 rounds of *I Spy* with someone at home.
Be sure to use defining attributes.
List the clues and shape below for one round.

MRB
125

Practice

② Write <, >, or =.

23 ☐ 32   13 + 5 ☐ 18   65 ☐ 43

Halves

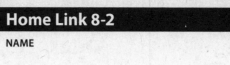

Family Note

Today your child began exploring fractions. Children divided circles and rectangles into two equal parts and identified and named halves.

Please return this Home Link to school tomorrow.

(1) Divide each square into 2 equal shares.
Try to divide each square in a different way.

(2) Name one of the shares.

Practice

(3) Use two of the digits in each group.
Write the smallest 2-digit number you can.

3, 1, 4, 5   9, 7, 8, 6

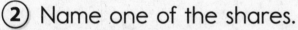

_____   _____

Fourths

Family Note

In the previous lesson, children divided objects into 2 equal shares. Today they divided circles and rectangles into 4 equal shares and discussed names for these shares. They also compared the sizes of the shares. They learned that larger wholes lead to larger shares, so half of a large pizza is larger than half of a small pizza. They also learned that having more shares means that each share will be smaller. So a pizza divided into 6 shares has larger shares than the same pizza divided into 8 shares.

Please return this Home Link to school tomorrow.

MRB
133

① Show 3 ways to divide the squares into 4 equal shares.

 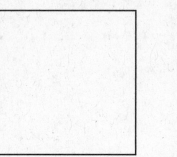 

② Name one of the parts.

Practice

③ How many more action figures are there than dolls?
_____ more action figures

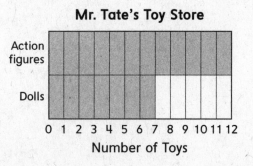

Mr. Tate's Toy Store

two hundred fifteen   215

More Equal Shares

MRB
132-133

① Pretend the circle is an orange slice. Divide it into two equal parts.

Name one of the parts.

_____

② Pretend the circle is another orange slice. Divide it into four equal parts.

Name one of the parts.

_____

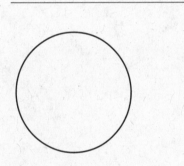

③ Which is bigger: 1 out of 2 equal parts of an orange slice, or 1 out of 4 equal parts?

_____

Tell someone at home how you know.

Practice

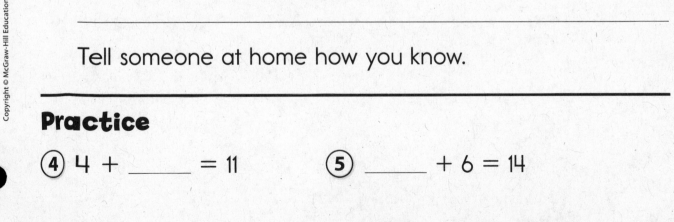

④ 4 + _____ = 11

⑤ _____ + 6 = 14

Composing Shapes

Family Note

Today your child used shapes such as triangles, squares, trapezoids, half circles, and quarter circles to make new shapes and designs.

Please return this Home Link to school tomorrow.

① Cut out the four shapes from the side of the page.

Use two or more shapes to fill one triangle. Trace around the pieces to show how they fit together.

Then use different shapes to fill the other triangle.

MRB
129

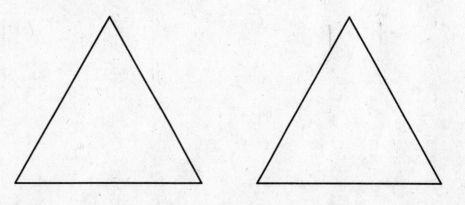

Practice

② Find 3 shapes that have 4 straight sides in your home. Draw them on the back of this page.

Defining Attributes of 3-Dimensional Shapes

Family Note

Throughout Unit 8, your child has been working with shapes. Today children learned about defining and nondefining attributes of 3-dimensional shapes.

Please return this Home Link to school tomorrow.

Find an object around the house that is shaped like a cube.

MRB
136

① What is the object? _____

② How many faces does it have? _____

③ What color is it? _____

④ What shape are its faces? _____

⑤ What is it made of? _____

⑥ Which answers are true for all cubes?

_____

Practice

⑦ Draw a polygon with 4 sides and 4 corners.

Composing Shapes and Fact Strategies

① How can you solve 6 + 7 using the near-doubles strategy?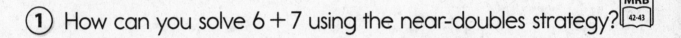

_____

_____

② How can you solve 6 + 7 using the making-10 strategy?

_____

_____

③ Which strategy is easier for you? Explain.

_____

_____

Practice

④ Pedro painted 12 pictures. He gave some away.
Now he has 5. How many pictures did Pedro give away?

_____ pictures   Number model: _____

Telling Time to the Half Hour

Family Note

Today your child began telling time to the nearest half hour on analog and digital clocks. Work together to complete these pages. Tell your child at which times, on the hour or half hour, he or she wakes up and goes to bed on school days. Have your child practice telling the time at home when it is close to the hour or half hour.

Please return these Home Link pages to school tomorrow.

MRB
106-107

Record the time.

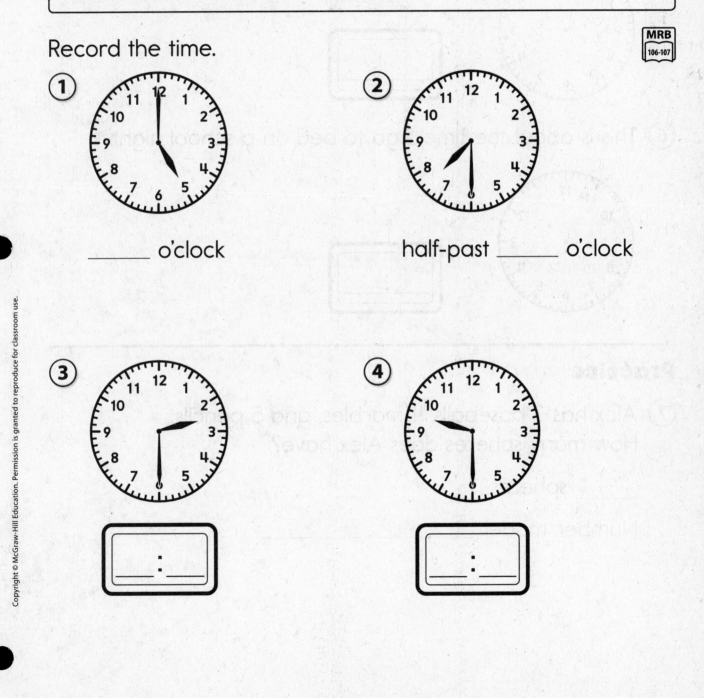

① _____ o'clock

② half-past _____ o'clock

③

④

Telling Time to the Half Hour (cont.)

Draw the hour hand and the minute hand to show the time.
Then write the time.

(5) This is about the time I wake up on a school day.

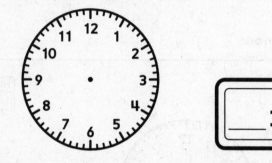

_____ : _____

(6) This is about the time I go to bed on a school night.

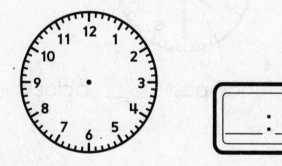

_____ : _____

Practice

(7) Alex has 3 baseballs, 4 marbles, and 5 pencils.
How many spheres does Alex have?

_____ spheres

Number model: _____

Review: Data

Family Note

Today your child's class took a survey. After everyone answered the survey, children totaled the results and displayed the data on a bar graph. The survey results below are similar to what children used in class today. Work with your child to make a bar graph and to answer the questions.

Please return this Home Link to school tomorrow.

A class took a survey about their favorite subjects.

MRB
116

① Complete the bar graph.

5 children liked Reading best.
6 children liked Math best.
2 children liked Spelling best.

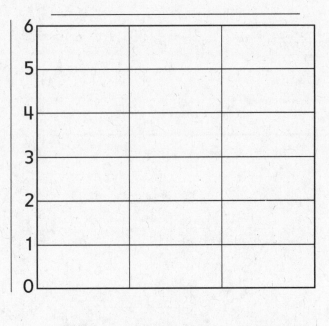

② How many children took the survey?

_____

③ How many more children chose Math than Spelling?

_____

Practice

④ Draw lines to divide the square into fourths.

two hundred twenty-seven   227

Number-Grid Puzzles

Family Note

Ask your child to show you how to complete the number-grid puzzles below. Encourage him or her to explain number-grid patterns that are helpful for solving the problems. For example, if you move up one row, the digit in the tens place is 1 less.

Please return this Home Link to school tomorrow.

Show someone at home how to fill in the missing numbers. **MRB 68**

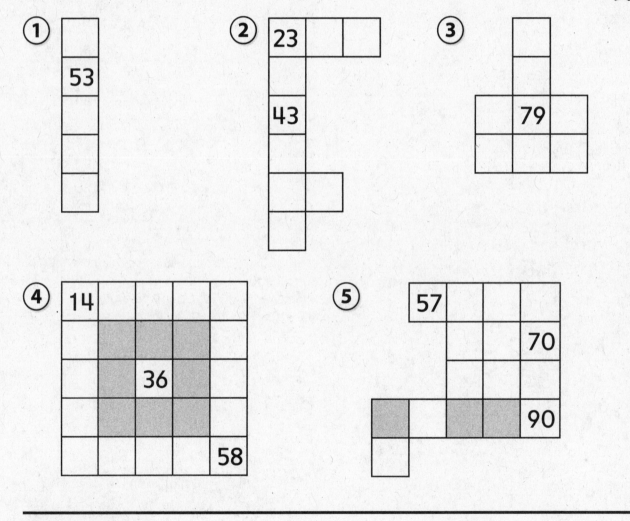

①
53

② 23
   43

③ 79

④ 14
   36
   58

⑤ 57
   70
   90

Practice

⑥ 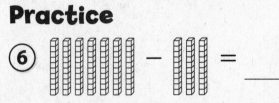 — = _____

Mentally Finding 10 More and 10 Less

Family Note

In earlier lessons, children used classroom tools, such as a number grid, base-10 blocks, and dimes, to help them add and subtract 10 from a given number. Today they made calculations mentally, using only their brains. Work with your child on adding and subtracting 10 mentally. Have your child do the routine below several times.

Please return this Home Link to school tomorrow.

Ask someone at home to say any number between 10 and 99. Record the number and fill in the blanks after it two times below.

① Number: _____

There are _____ tens and _____ ones in _____.

10 more than _____ is _____. 10 less than _____ is _____.

② Number: _____

There are _____ tens and _____ ones in _____.

10 more than _____ is _____. 10 less than _____ is _____.

Practice

③ Elaine's farm has 4 cows, 2 goats, and 8 chickens. How many animals are there all together?

_____ animals   _____ + _____ + _____ = _____

Unit 9: Family Letter

Two-Digit Addition and Subtraction and Review

In Unit 9, children solve real-world problems about comparing prices and buying items from a school store or vending machine. They also solve silly number stories about animals. To solve these problems, children add and subtract pairs of 1-digit numbers, decade numbers (such as 40, 50 or 80), and 2-digit numbers. After solving the problems, they compare numbers using the symbols $<$, $>$, and $=$.

Children continue to work on measurement concepts by using paper clips to measure objects and creating a paper-clip ruler to measure more efficiently. *Second Grade Everyday Mathematics* builds on this by introducing rulers with inches and centimeters as units.

Children also find equivalent names for numbers by solving broken-calculator puzzles. This activity requires children to determine how to display numbers when certain calculator keys do not work.

Example: Imagine your 3-key is broken.

How can you show the number 13 without using the 3-key?

$9 + 4$ or $15 - 2$

Children also review other topics from *First Grade Everyday Mathematics* in Unit 9, including place value and geometry.

Do-Anytime Activities

To work with your child on the concepts taught in this unit and in previous units, try these activities:

1. Use Fact Triangles and any of the games introduced at school to help your child practice addition and subtraction facts.

2. Say a 2-digit number. Ask your child to mentally find 10 more and 10 less.

3. Have your child tell time to the hour or half-hour.

4. Find and describe geometric shapes in everyday objects with your child.

5. Have your child create the largest and smallest numbers possible when given 2 (or 3) digits.

6. Make up and solve broken-calculator puzzles.

Building Skills through Games

Your child will play these games and others in Unit 9:

Beat the Calculator

A "Calculator" (a player who uses a calculator) and a "Brain" (a player who does not use a calculator) race to see who will be first to solve addition facts.

Time Match

Players find pairs of cards showing matching times on analog and digital clocks like in the game *Concentration* (also known as *Memory*). The player with the most matching cards wins.

As You Help Your Child With Homework

As your child brings home assignments, you may want to go over the instructions together, clarifying them as necessary. The answers listed below will guide you through the Home Links for this unit.

Home Link 9-1

1–5. Answers vary.

6. <; =; >; <

Home Link 9-2

1. Answers vary.

2. ; 9:00

Home Link 9-3

1. 61

2. 7

Home Link 9-4

1. Sample answers: $20 + 10 =$; $10 + 10 + 10 =$; $29 + 1 =$

2. Sample answers: $8 + 7 =$; $14 + 1 =$; $16 - 1 =$

3. Sample answers: $9 + 9 =$; $20 - 2 =$; $8 + 8 + 2 =$

4. Sample answer:

| | | | | | | | | ▪ ▪ ▪ ▪ ▪ ▪ ▪ ▪ ▪ ▪ ▪ ▪ ▪ ▪

Home Link 9-5

1. 95 cents

2. 95 cents

3. Answers vary.

Home Link 9-6

1. No

2. 2¢; Sample answers: $44 + 2 = 46$; $46 - 44 = 2$

3. 20¢; Sample answers: $26 + 20 = 46$; $46 - 26 = 20$

4. 70¢; $44 + 26 = 70$

5. 24¢; Sample answers: $70 - 46 = 24$; $46 + 24 = 70$

6. 70; 22; 33

Home Link 9-7

1. 67; $47 + 20 = 67$

2. 74; $37 + 37 = 74$

3. 58; $22 + 26 + 10 = 58$

4. 50; 60; 30; 40

Home Link 9-8

1. rubber bands and box of crayons;
   Sample answer: 56 < 88

2. eraser and ball; Sample answer: 62 < 67

3. Smaller

Home Link 9-9

1.
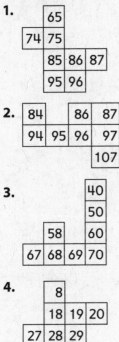

	65		
74	75		
	85	86	87
	95	96	

2.

84		86	87
94	95	96	97
			107

3.

			40
			50
		58	60
67	68	69	70

4.

	8	
	18	19	20
27	28	29	
	39	

5. 4; 6; 40; 60

Home Link 9-10

1. Cube

2. Cylinder

3. Rectangular prism or cube

4. Pyramid

5. Cone or cylinder

6. 13 ribbons; 6 + 4 + 3 = 13

Home Link 9-11

1.
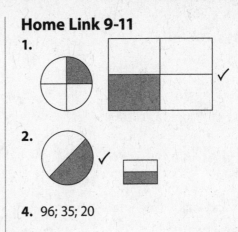

2.

4. 96; 35; 20

Reviewing Measurement

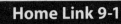

Family Note

Today your child reviewed measurement ideas learned in first grade. Children measured length in paper clips, then made rulers with paper clips as the units. Work with your child to select and measure items around the home using the paper-clip ruler. Please make sure your child brings his or her ruler back to class, as it will be used again.

Please return this Home Link to school tomorrow.

Find 5 paths or objects in your home to measure with your paper-clip ruler.

MRB
98

① I measured _____.

It is about _____ paper-clip units long.

② I measured _____.

It is about _____ paper-clip units long.

③ I measured _____.

It is about _____ paper-clip units long.

④ I measured _____.

It is about _____ paper-clip units long.

⑤ I measured _____.

It is about _____ paper-clip units long.

Practice

⑥ Write <, >, or =.

15 ____ 51   80 ____ 80   49 ____ 44   16 ____ 106

two hundred thirty-seven   237

2-Digit Number Stories

Family Note

Today your child practiced adding and subtracting 2-digit numbers by pretending to shop at a school store.

Sample Story

I bought a ball and an eraser. I paid 52¢. Number model: *35¢ + 17¢ = 52¢*

Please return this Home Link to school tomorrow.

(1) Think of two things to buy. Draw a picture and write a number story about buying them. Use the back of the page if needed. Then write a number model.

MRB 24-26

Number model: _____

Practice

(2) Record the time.

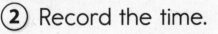

_____ : _____

Shopping for School Supplies

Family Note

Today your child practiced explaining solution strategies clearly. Clear explanations make sense to the listener and include all of the steps used to solve the problem. For example, "I started at 21 and counted up," would not be a clear explanation for how a child added 21 and 7. An example of a clear explanation would be, "I started at 21. Then I counted up 7. I ended at 28. So, 21 + 7 = 28."

After your child solves the problem below, ask him or her to explain the strategy used. Ask questions to encourage your child to explain the strategy clearly.

Please return this Home Link to school tomorrow.

① You bought these items at the school store. How much did you pay in all?

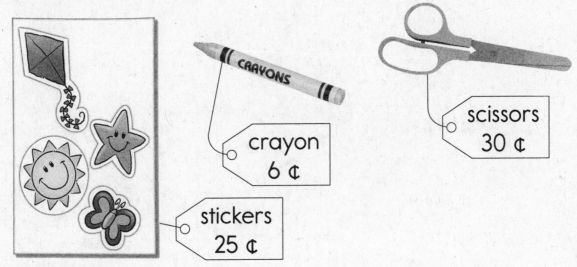

crayon
6 ¢

scissors
30 ¢

stickers
25 ¢

I paid _____ cents.

Explain to someone at home how you solved the problem.

Practice

② Solve.

$20 = 8 +$ _____ $+ 5$ $17 = 6 +$ _____ $+ 7$

two hundred forty-one   241

Broken-Calculator Puzzles

Use + and − to solve the broken-calculator puzzles.
Use a calculator to check your answers.

① Imagine your 3-key is broken.
Write at least three ways to show 30.

② Imagine your 5-key is broken.
Write at least three ways to show 15.
Use subtraction in one of the ways.

③ Imagine your 1-key is broken.
Write at least three ways to show 18.

Practice

④ Jake has | | | | | | | | | and ▪ ▪▪

Show one exchange he could make.

Vending Machine Addition and Subtraction

Family Note

Today your child added and subtracted prices of items found in a vending machine. Ask your child to explain how to solve Problems 1 and 2 below. Provide pennies and dimes to help your child model each problem.

Please return this Home Link to school tomorrow.

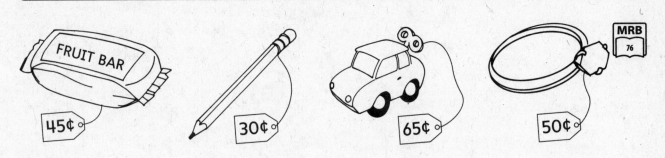

FRUIT BAR   45¢   30¢   65¢   50¢   MRB 76

Solve.

① How much does it cost to buy a pencil and a toy car?

_____

② How much does it cost to buy a fruit bar and a ring?

_____

Practice

③ Put together shapes to make 2 new shapes on the back of this paper.
Use a triangle, square, and a half circle.

More 2-Digit Number Stories

Solve the problems and write the number models.
Ramona has 44¢. Scott has 26¢. A stamp costs 46¢.

MRB
82-83

(1) Can Ramona or Scott buy a stamp? _____

(2) How much more money does Ramona need? _____ ¢

Number model: _____

(3) How much more money does Scott need? _____ ¢

Number model: _____

(4) How much money do Ramona and Scott have all together? _____ ¢

Number model: _____

(5) If they buy a stamp together, how much money will they have left? _____ ¢

Number model: _____

Practice

(6) Solve.

$30 +$ _____ $= 100$ $67 =$ _____ $+ 45$ $12 + 21 =$ _____

Strategies for 2-Digit Addition

Family Note

Today your child continued using various strategies to solve number stories involving adding and subtracting larger numbers. Encourage your child to explain how more than one strategy can be used to solve each of the problems on this page.

Please return this Home Link to school tomorrow.

MRB
90

Solve.

Write a number model for each story.

(1) Daniel built a tower with blocks.
It had 47 cubes and 20 cylinders.
How many blocks are in Daniel's tower? _____ blocks

Number model: _____

(2) Carmen used blocks to make a fort.
She used 37 cubes and 37 cones.
How many blocks are in her fort? _____ blocks

Number model: _____

(3) Janet built a tower out of blocks.
She used 22 cubes, 26 cylinders, and 10 cones.
How many blocks are in Janet's tower? _____ blocks

Number model: _____ + _____ + _____ = _____

Practice

(4) Solve. Use dimes if you like.

90 − 40 = _____   80 − 20 = _____

60 − 30 = _____   70 − 30 = _____

Review: Relations and Equivalence

Ryan and Janae are choosing things to buy.
Circle the group that costs more money.

Write a number model with < or > to compare the prices.

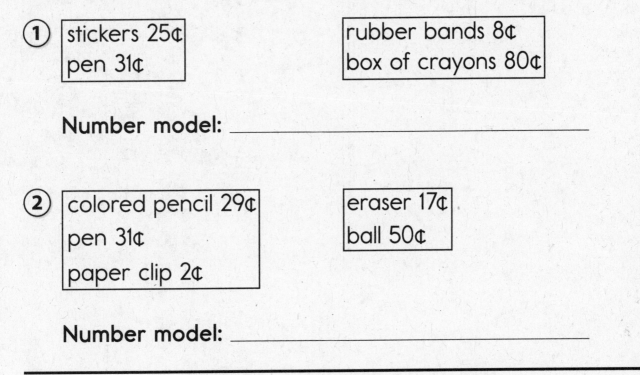

① | stickers 25¢
   | pen 31¢

   rubber bands 8¢
   box of crayons 80¢

Number model: _____

② | colored pencil 29¢
   | pen 31¢
   | paper clip 2¢

   eraser 17¢
   ball 50¢

Number model: _____

Practice

③ Jada and Martin cut a pizza in half to share.
Then Min and Julius want to share the pizza, too.
So they cut the pizza into fourths.

Are the shares now larger or smaller? _____

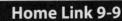
Family Note

Today your child reviewed place value. Children also completed number-grid puzzles for 2-digit numbers. Ask your child to explain how to solve each problem below.

Please return this Home Link to school tomorrow.

Fill in the missing numbers.

MRB
67-68

①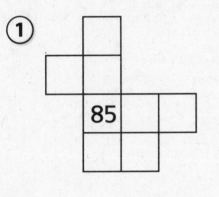

②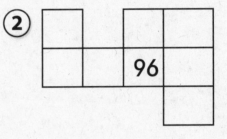

③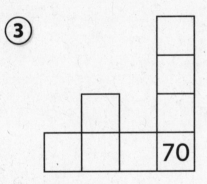

④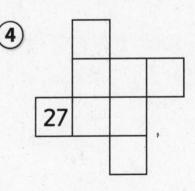

Practice

⑤ Subtract.

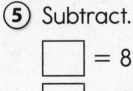 = 8 − 4

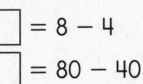 = 80 − 40

9 − 3 = ☐

90 − 30 = ☐

Review: 3-Dimensional Geometry

Family Note

Today your child reviewed attributes of 3-dimensional shapes. Ask your child to point out objects of various shapes around your home or outside and name their defining attributes.

Please return this Home Link to school tomorrow.

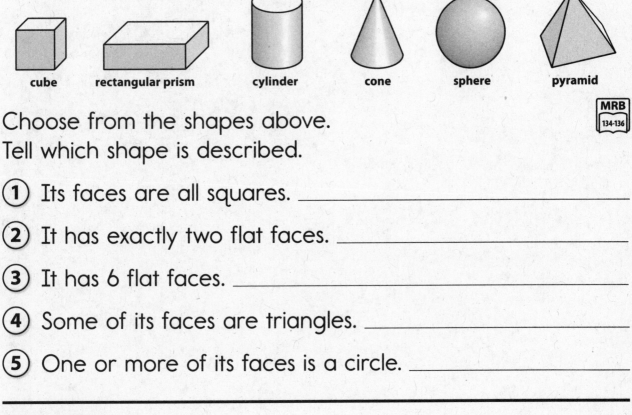

cube   rectangular prism   cylinder   cone   sphere   pyramid

Choose from the shapes above.
Tell which shape is described.

MRB
134-136

① Its faces are all squares. _____

② It has exactly two flat faces. _____

③ It has 6 flat faces. _____

④ Some of its faces are triangles. _____

⑤ One or more of its faces is a circle. _____

Practice

⑥ Risa has 6 red ribbons, 4 blue ribbons, and
3 purple ribbons.

How many ribbons does she have? _____

Number model: _____ + _____ + _____ = _____

Review: Equal Shares

Family Note

Today your child reviewed dividing rectangles and circles into 2 or 4 equal shares. Children were reminded about the different names for these shares and for the whole. Help your child divide the shapes below and compare the sizes of the shares.

Please return this Home Link to school tomorrow.

① Divide each shape in fourths. Shade 1 fourth.

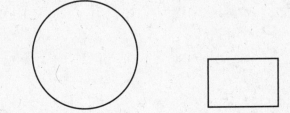

② Divide each shape in half. Shade 1 half.

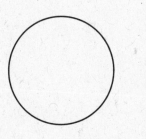

③ Put a check next to the circle with the largest shares.
Put a check next to the rectangle with the largest shares.

Practice

④ Solve. Use pennies and dimes to help you.

$86 + 10 =$ _____

_____ $= 45 - 10$

$70 - 50 =$ _____

End-of-Year Family Letter

Congratulations! By completing *First Grade Everyday Mathematics* your child has accomplished a great deal. Thank you for all of your support.

This Family Letter is provided for you to use as a resource throughout your child's school vacation. It includes a list of Do-Anytime Activities, game directions, fact practice tips, and a sneak preview of what your child will be learning in *Second Grade Everyday Mathematics*.

Enjoy your summer!

Do-Anytime Activities

The following activities are for you and your child to do together over the summer to help review concepts your child learned in first grade. These activities build on the skills from this year and help prepare him or her for *Second Grade Everyday Mathematics*.

Telling Time and Measuring Length

- Tell time to the hour and half hour on analog and digital clocks in a variety of situations.

- Set alarm clocks and timers on objects such as ovens, microwave ovens, and mobile phones.

- Record the time spent doing various activities.

- Measure lengths of objects and paths with nonstandard units such as paper clips, toothpicks, or arm spans.

Collecting Data

- Collect data by asking questions:

    Which is your favorite summer fruit—watermelons, strawberries, or peaches?

- Collect data by making observations:

    How many people are wearing shorts, dresses, or swimsuits?

- Organize data in tally charts and in bar graphs, including keeping track of the weather.

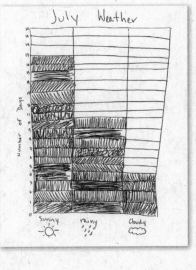